AGRICULTURAL ALMANAC,

FOR THE YEAR OF OUR LORD

1984,

Being a Bissextile or Leap Year, until the 4th of July, the 208th of American Independence

CONTAINING

The rising, setting, and eclipses of the Sun and Moon; the phases and places of the Moon; the aspects of the planets, the rising, setting, and southing of the most conspicuous planets and fixed stars, and equation of time, with a variety of useful and entertaining matter.

Carefully Calculated for the Meridian of Pennsylvania and the adjoining States by HART WRIGHT COMPANY, Lewiston, Maine, with Weather Predictions, Sun and Moon Tables, and additional astronomical information for all parts of the nation. Inquiries about astronomical data should be mailed direct to Hart Wright Company.

Published by Golden Press/New York Western Publishing Company, Inc./Racine, Wisconsin
ISSN 0732-5932
ISBN 0-307-46651-5
Produced in USA
Golden® and Golden Press® are trademarks of Western Publishing Company, Inc.

Published simultaneously in Canada

How to Use the Almanac and Understand It

An almanac has a variety of uses, some of which are easy to understand such as advice on planting flowers, others more difficult such as the astronomical tables for the months.

This is an effort to tell you what you can find, what it can mean to you, and how to make it useful.

On page 6 you will find symbols which are part of astronomical shorthand. There are 12 zodiacal constellations, each represented by one sign of the Zodiac, an imaginary belt in the skies which has 12 divisions. Page 6 also presents the Moon signs and other symbols with meanings of their own.

All these signs appear in the astronomical, left-hand, page for each month. The Zodiac signs, under Place & Age, tell you in which constellation the Moon will be each day. The Miscellaneous Particulars column contains a variety of sky information. To learn more from this column, read a basic text on astronomy. You will find it fascinating. Aldebaran, for example, is a red star in the eye of the constellation Taurus, and the brightest star in the Hyades. When the Hyades rose with the sun, the ancients said they foretold rain.

Here are the explanations of what you will find in other columns:

Moon South — The figure denotes the time the Moon crosses the meridian.

Moon Place — The figures give age of the moon each day.

Moon Rises (or Sets) — This tells you the hour, and you can determine whether it is a.m. or p.m.

Moon's Phases — These are in Eastern Standard Time.

Time — The Sun and Moon Rises are given in Civil or mean time, based on the hour and the minute of the 75° W, or the central meridian of your time zone.

A correction for your longitude is necessary. The correction is the number of minutes (1° = 4 minutes of time) of time that the place is either west (plus) or east (minus) of the standard meridian of your time zone. The standard meridian for those keeping Eastern Standard Time is the 75° meridian. The longitude of Philadelphia is nearly 75° W, so no correction is necessary there.

The Rising, Setting and Southings or Meridian Passage for stars and constellations are given in Civil Time.

The Sun — Slow or Fast indicates the number of minutes that local mean time is ahead of or behind sundial or apparent time. The difference can range from 0 to 16 minutes. The Rises h. m. column, for the Sun, tells you when the sun appears to rise over the horizon, or when it appears to set below the horizon. There may be light in the sky before sunrise, and light after sunset.

Ember Days — In the Roman Catholic and Anglican churches, days set for fasting and prayer in each season.

Dominical Letter — If New Year's Day falls on a Sunday, the Dominical Letter is A, and from it you can tell the day of the week on which any date falls. If New Year's Day is a Monday, the Dominical Letter is B, and so on.

Epact — The Moon's age at the beginning of the year; useful in finding the Paschal Full Moon, from which the date of Easter is determined.

Lunar Cycle or Golden Number — This is a period of 19 years, after which time the New and Full Moon fall approximately on the same day of the month as they did at the beginning of the period. Any number of this period is called the Golden Number. To find the Golden Number, add 1 to the given year and divide by 19. The remainder is the Golden Number. If the remainder is 0, the Golden Number is 19. The Golden Number is also listed on page 6.

Solar Cycle — A period of 28 years, after which the days of the month return to the same days of the week. Its chief use is to find the Dominical Letter.

Roman Induction — A cycle of 15 years, reckoned from the year as its origin. Hence, if any given year less 313 be divided by 15 the remainder will be the current year of the period.

Julian Period — Reckoned from the year 4713 B.C., and may be found by adding 4713 to any given year. Its chief use is found in chronological data.

Quick Reference Mileage Chart

	ALBUQUERQUE	ATLANTA	BOSTON	CHICAGO	CLEVELAND	DALLAS	DENVER	DETROIT	LOS ANGELES	MIAMI	MINNEAPOLIS	NEW ORLEANS	NEW YORK CITY	ST. LOUIS	SALT LAKE CITY	SAN FRANCISCO	SEATTLE	WASHINGTON, D.C.
ALBUQUERQUE		1403	2266	1350	1612	651	421	1595	830	1981	1276	1155	2043	1058	625	1138	1487	1894
ATLANTA	1403		1079	708	725	811	1428	738	2216	675	1115	500	860	565	1935	2553	2874	633
BOSTON	2266	1079		990	646	1814	1998	711	3092	1543	1397	1549	219	1208	2407	3188	3066	445
CHICAGO	1350	708	990		344	925	1008	279	2166	1383	407	976	840	292	1439	2198	2076	702
CLEVELAND	1612	725	646	344		1195	1339	177	2450	1400	751	1096	509	554	1774	2542	2420	370
DALLAS	651	811	1814	925	1195		780	1171	1405	1330	966	504	1595	633	1260	1789	2150	1371
DENVER	421	1428	1998	1008	1339	780		1287	1132	2086	855	1282	1848	865	507	1266	1370	1703
DETROIT	1595	738	711	279	177	1171	1287		2425	1408	686	1058	637	537	1718	2477	2357	535
LOS ANGELES	830	2216	3092	2166	2450	1405	1132	2425		2735	1972	1909	2873	1888	730	412	1143	2720
MIAMI	1981	675	1543	1383	1400	1330	2086	1408	2735		1790	877	1331	1235	2593	3119	3455	1107
MINNEAPOLIS	1276	1115	1397	407	751	966	855	686	1972	1790		1251	1247	567	1245	2004	1669	1109
NEW ORLEANS	1155	500	1549	976	1096	504	1282	1058	1909	877	1251		1330	684	1764	2283	2626	1106
NEW YORK CITY	2043	860	219	840	509	1595	1848	637	2873	1331	1247	1330		985	2279	3038	2916	224
ST. LOUIS	1058	565	1208	292	554	633	865	537	1888	1235	567	684	985		1370	2134	2226	840
SALT LAKE CITY	625	1935	2407	1439	1774	1260	507	1718	730	2593	1245	1764	2279	1370		759	864	2128
SAN FRANCISCO	1138	2553	3188	2198	2542	1789	1266	2477	412	3119	2004	2283	3038	2134	759		821	2887
SEATTLE	1487	2874	3066	2076	2420	2150	1370	2357	1143	3455	1669	2626	2916	2226	864	821		2778
WASHINGTON, D.C.	1894	633	445	702	370	1371	1703	535	2720	1107	1109	1106	224	840	2128	2887	2778	

CHARACTERS OF THE CONSTELLATIONS

♈ Aries, the Ram	♌ Leo, the Lion	♐ Sagittarius, the Bowman
♉ Taurus, the Bull	♍ Virgo, the Virgin	♑ Capricorn, the Goat
♊ Gemini, the Twins	♎ Libra, the Balance	♒ Aquarius, the Waterman
♋ Cancer, the Crab	♏ Scorpio, the Scorpion	♓ Pisces, the Fishes

ASTRONOMICAL CHARACTERS EXPLAINED

New Moon.

First quarter, or Moon in general.

Full Moon.

Last quarter, or Moon in general.

☊ Moon's Ascending Node, or Dragon's Head.

☋ Moon's Descending Node, or Dragon's Tail.

◡ Moon's Ascension.

◠ Moon's Descension.

☽ Moon in Apogee, furthest from the Earth.

☾ Moon in Perigee, nearest to Earth.

PLANETS AND ASPECTS

☉ Sun.
☿ Mercury.
♀ Venus.
♆ Neptune.
⊕ Earth.
♂ Mars.

♃ Jupiter.
♄ Saturn.
♅ Herschel, or
⛢ Uranus.
☾ Moon.

☌ Conjunction, or planets in the same longitude.

□ Quartile, when they are 90 degrees apart.

☍ Opposition, when they are 180 degrees apart.

NOTE: The Sun and Moon Rises and Sets are given in Civil Time. The Moon's Phases are given in Eastern Standard Time and the Rising, Setting and Southing or Meridian Passage of the Stars, Constellations or Signs, in Civil Time. For the Standard Time applicable to Lancaster and vicinity add 4 minutes to the Civil Time. For Daylight Saving Time add an additional hour. The Moon's Place is given for 7:00 A.M. Eastern Standard Time.

MOVABLE FEASTS IN 1984

Baptism of Jesus	Jan. 15
Shrove Tuesday	Mar 6
Ash Wednesday	Mar. 7
First Sunday in Lent	Mar. 11
Palm Sunday	Apr. 15
Maundy Thursday	Apr. 19
Good Friday	Apr. 20
Easter	Apr. 22
Mother's Day	May 13
Rogation Sunday	May 27
Ascension Day	May 31
Pentecost — Whitsunday	June 10
Trinity Sunday	June 17
Father's Day	June 17
Labor Day	Sept. 3
Gold Star Mothers' Day	Sept. 30
Election Day	Nov. 6
Thanksgiving Day	Nov. 22
Advent Sunday	Dec. 2
Sundays after Pentecost this year	24

MERCURY is the "Reigning Planet" this year.

QUATEMBER OR EMBER DAYS

1st, March 14, 16, 17
2nd, June 13, 15, 16
3rd, September 19, 21, 22
4th, December 19, 21, 22

CHRONOLOGICAL CYCLES

Dominical Letter	AG
Epact	27
Lunar Cycle or Golden Number	9
Solar Cycle	5
Roman Indiction	7
Julian Period	6697
Dionysian Period	313
Jewish Lunar Cycle	6

The chronology of the Jews for the year 5745 begins at sunset September 27, 1984.

The chronology of the Mohammedans for the year 1405 begins at sunset September 27, 1984.

Eclipses for the Year 1984

Eastern Standard Time

In 1984 there will be five eclipses, two of the sun and three of the moon. All three lunar eclipses will be penumbral, hence, uninteresting. No part of the moon will enter the deep, inner shadow of the earth; thus, there is little to look at. They are included here for completeness of information.

I. May 14 and 15, a penumbral eclipse of the moon. The beginning is visible in South America, Antarctica, Africa, except the extreme eastern part, Madagascar, Europe, except the northeastern part, North America, except the northwestern part, Southern Greenland, the Atlantic Ocean, and the eastern part of the Pacific Ocean. The end is visible in South America, North America, except the extreme northern part, Antarctica, New Zealand, the extreme western part of Africa, the Atlantic Ocean, and the Pacific Ocean.

The eclipse will be visible throughout most of North America, beginning May 14, 9:42 p.m., mid-eclipse 11:40 p.m., ending at 1:38 a.m., May 15.

II. May 30, an annular eclipse of the sun. The partial phase, when the moon covers only part of the sun, will be visible throughout Central and North America, except Alaska, extreme northern South America, Greenland, Europe, and western Africa. The path of annularity passes over Mexico and the United States from approximately New Orleans, Louisiana to northern Assateague Island, south of Ocean City, Maryland. Mid-eclipse occurs at 11:53 a.m.

III. June 13, a penumbral eclipse of the moon. This eclipse is invisible in the western hemisphere and is invisible where it's visible — the Antipodes — for less than 10 percent of the diameter of the moon will enter the penumbra of the earth. Totally unnoteworthy.

IV. November 8, a penumbral eclipse of the moon. This eclipse is also not visible in the western hemisphere, but the other side of the world will see a slight darkening of the northern part of the moon.

V. November 22 and 23, a total eclipse of the sun, invisible in North America. The path of totality begins west of New Guinea, passes the length of southern New Guinea, north of New Zealand, and across the Pacific Ocean.

Centennial Almanac for the Year 1984

The following is an extract from the Centennial Almanac for 1984

MERCURY is the Reigning Planet this year.

YEAR IN GENERAL. Is more dry and cold than warm, and seldom fruitful.

SPRING. The end of March will be warm, April until the 25th, dry, afterwards cold; May will have some cold days in the commencement, so that the fruit will be in danger.

SUMMER. Will have a good deal of rain, by which the earth will however not be properly refreshed. The hay and grain can be brought in pretty well, though it will be best not to delay with it.

AUTUMN. In the commencement there will be a great deal of rain, and there will be an early frost; but when the half of October is past, there will be dry weather, until the commencement of Advent.

WINTER. After a pleasant autumn, winter will fall in the commencement of December; it will be cold and snow until February, which will appear rather mild; but from the middle until the 4th of March it will be cold, and from thence until the termination storms and tempests will take place.

CULTIVATION OF SUMMER GRAIN. Will succeed pretty well. It is a favorable year for barley and oats; peas and other leguminous plants must be sown when the ground is neither too dry nor too wet and damp. Hemp and flax will be good, but the hemp will remain short.

CULTIVATION OF WINTER GRAIN. Is various. If the summer of the preceding year was very warm, wheat and rye will in the present year yield largely in straw, but little in the grain; but if the preceding year was humid, they will yield well in the grain. Rye and wheat will be endangered in the blossoming.

AUTUMN SEEDING. The early and the late will be the best; the middle will be consumed by the flies. The early sown grain can be pastured to the sheep in the spring, but the late sown not.

FRUIT. In some places it will grow abundantly, in some places tolerably well, and in others there will be none at all.

HOPS. Will not be abundant nor strong.

TEMPESTS and THUNDERSTORMS. The east wind will be the most prevalent, sometimes the west but seldom the north wind; during the summer there will be but few thunderstorms.

DISEASES. Can be cured only slowly in this year, and they will prevail mostly in the spring and the latter part of autumn. Epidemic diseases generally prevail among the cattle this year.

TABLE OF CONTENTS

Significant Days for 1984

New Year's Day	January 1
Epiphany (Episcopal)	January 6
Epiphany (Catholic)	January 8
Baptism of Jesus	January 15
Martin Luther King's Birthday	January 15
Robert E. Lee's Birthday	January 19
F. D. Roosevelt's Birthday	January 30
Purification (Candlemas)	February 2
Groundhog Day	February 2
Boy Scouts Day	February 8
Lincoln's Birthday	February 12
St. Valentine's Day	February 14
Washington's Birthday (observed)	February 20
Shrove Tuesday	March 6
Mardi Gras	March 6
Ash Wednesday	March 7
First Sunday in Lent	March 11
Girl Scouts Day	March 12
Fast of Esther	March 15
St. Patrick's Day	March 17
Purim	March 18
Spring Begins	March 20
Annunciation Day	March 25
Thomas Jefferson's Birthday	April 13
Pan-American Day	April 14
Palm Sunday	April 15
Patriots' Day	April 16
First Day of Passover	April 17
Maundy Thursday	April 19
Good Friday	April 20
Easter	April 22
Daylight Saving Time Begins	April 29
Mother's Day	May 13
Peace Day	May 18
Armed Forces Day	May 19
Lag B'Omer	May 20
National Maritime Day	May 22
Rogation Sunday	May 27
Memorial Day	May 28
Ascension Day	May 31
Greek Orthodox Ascension Day	May 31
Jeff Davis' Birthday	June 3
Shavuot	June 6
Pentecost — Whitsunday	June 10
Children's Day	June 10
Greek Orthodox Pentecost	June 10
Flag Day	June 14
Trinity Sunday	June 17
Father's Day	June 17
Corpus Christi	June 21
Summer Begins	June 21
St. John The Baptist (Midsummer Day)	June 24
Dog Days Begin	July 3
Independence Day	July 4
St. Swithin's Day	July 15
Fast of Tammuz	July 17
Fast of Av	August 7
Dog Days End	August 11
Assumption B.V.M.	August 15
Labor Day	September 3
Grandparents' Day	September 9
Harvest Moon	September 10
Holy Cross Day	September 14
Citizenship Day	September 17
Autumn Begins	September 22
Rosh Hashanah (New Year)	September 27
Fast of Gedaliah	September 30
Gold Star Mothers' Day	September 30
Yom Kippur	October 6
Columbus Day (observed)	October 8
Fire Prevention Day	October 9
Succot	October 11
Simchat Torah	October 19
United Nations Day	October 24
Daylight Saving Time Ends	October 28
Halloween	October 31
All Saints' Day	November 1
Election Day	November 6
Veterans' Day	November 11
Thanksgiving Day	November 22
Advent Sunday	December 2
Conception B.V.M.	December 8
Chanukah	December 19
Winter Begins	December 21
Christmas Day	December 25
New Year's Eve	December 31

ANATOMY OF MAN'S BODY, AS SAID TO BE GOVERNED BY THE TWELVE CONSTELLATIONS

The Head and Face — Aries.

Arms, Gemini.

Heart, Leo.

Reins, Libra.

Thighs, Sagittarius.

Legs, Aquarius.

Neck, Taurus.

Breast, Cancer.

Bowels, Virgo.

Secrets, Scorpio.

Knees, Capricorn.

The Feet — Pisces.

Presidential Succession:

By Richard A. Snyder

The law is the law — but is the law sufficient to meet all situations which may arise? In this presidential election year, the question should be addressed. This matter, of national importance, is aptly discussed here by Sen. Richard A. Snyder, chairman of the Judiciary Committee of the Pennsylvania State Senate, analyst of government, and skillful writer.

Who takes over when the President dies? The Vice President, of course. Who fills in for the President when he is just incapacitated? Good question. Sometimes his wife, his secretary or friends. It all depends. At any rate, the scenario cannot be predicted.

Strange situations have arisen, especially in the cases of Presidents Woodrow Wilson and James A. Garfield, both of whom were unable to fulfill their duties over extended periods. The two most remembered martyrs, Abraham Lincoln and John Kennedy, survived for so short a time after being attacked that the interval posed no problem.

Some mystery shrouds the situation of President Wilson who suffered a stroke in October 1919. It would appear that Mrs. Wilson, a strong personality and his second wife, took charge, made the decisions, prevented officials from seeing the patient, and kept the situation so ambiguous that there was no effort to formalize a substitute.

William Henry Harrison in 1841 was the first president to suffer a fatal illness in office. He caught a cold giving his address at a blustery inaugural March 4, developed pneumonia, was bedfast March 27, and died April 4, serving so brief a time and in so simple an era that no interruption in presidential duties was of national concern.

Grover Cleveland suffered cancer of the roof of his mouth. This was kept secret by collaboration of his doctors, his secretary and other intimates. Because the economy was precarious and panic was feared, the President was taken aboard a yacht in New York's East River in 1893, where the operation was performed, successfully. It was only years afterwards that the matter became public knowledge.

President McKinley died eight days after an assassin shot him on September 6, 1901. The brevity of his confinement in a time of tranquility posed no national problems, although Vice President Theodore Roosevelt, taking office, made substantial changes in policy and changed the nature of the presidential office.

James A. Garfield was assassinated July 2, 1881 and his lingering situation, until his death 79 days later, created perhaps the most unusual one in the history of the presidents. Garfield had difficulty in choosing a private secretary from among several mature persons. Meanwhile, he temporarily used the services of a 22-year old, who was skilled at stenography and had a gift for drafting prose in Garfield's style.

President Garfield **President Wilson**

The secretary was Joseph Stanley Brown, who is described by Michael Medved in his volume "The Shadow Presidents," as gangling and homely, and it was assumed Brown was beyond his depth as an aide to a robust president, and even less qualified to aid a stricken one. Yet in the days which followed, until Garfield's death, the young man performed admirably. Brown had the unique ability to answer a letter as a healthy Garfield would have answered. In addition, he defended the President from visitors, and also served to help the Garfield family through the President's illness and after his passing.

He was rewarded. The Garfields regarded him as one of their family, helped him through Yale. He subsequently married the President's daughter, Mollie. He went into geology and railroading and lived to become wealthy, but no later episodes were quite as exciting as the

'Solution' Leaves Twilight Zone

months he spent as a virtual stand-in for the President.

A formalized substitution for a healthy president occurred when Dwight Eisenhower suffered a coronary thrombosis in 1955 and, with his approval, Vice President Nixon and the cabinet met on the normal schedule and transacted all government business except that which required the President's specific approval. Eisenhower's two subsequent illnesses in office, the inflammation of his small intestine in 1956 and a mild stroke in 1957, were handled similarly.

The Eisenhower-Nixon example paved the way for the 25th amendment to the Constitution, regarded as a "solution" for cases in which either the President is willing to yield up his power temporarily to the Vice President or, when unable to do so, his powers can be assumed by the Vice President and the Cabinet by initiative on their part. This contemplates that the President resumes his duties when able.

A twilight zone of circumstances remains, however, a fact which Eisenhower has referred to in his memoirs. There can be situations in which a solution is elusive. The human body is vulnerable in many ways and it may occur that a half-stricken executive will not agree to be replaced. A conflict for the reins of government may then occur. The fears are augmented by the fact that most presidents are men of middle age or older, and the pressures of the office tax even a sturdy physique.

The public's knowledge of the condition of the president also showed an abrupt change about 30 years ago. Eisenhower took the position that the public was entitled to know the precise medical condition of their president, although he was was somewhat embarrassed by details which were published. His doctor, Paul Dudley White, heart specialist, justified this on the grounds that by publishing details, doctors all over the nation would understand more of his precise condition and be reassured.

From the standpoint of the nation, a president of deteriorating capacity may be unable to promote his programs, as Wilson was unable to do with his League of Nations in his last year, 1920. The decisions made at Yalta on Roosevelt's final trip were also criticized later as ones which would not have been made had he been in robust health. Roosevelt died two months after he returned from Yalta.

PRESIDENTS OF THE UNITED STATES

Name	Politics	Native State	Inaugurated	Age	Age at Death
1. George Washington	F	Va.	1789	57	67
2. John Adams	F	Mass.	1797	61	90
3. Thomas Jefferson	D-R	Va.	1801	57	83
4. James Madison	D-R	Va.	1809	57	85
5. James Monroe	D-R	Va.	1817	58	73
6. John Quincy Adams	D-R	Mass.	1825	57	80
7. Andrew Jackson	D	S.C.	1829	61	78
8. Martin Van Buren	D	N.Y.	1837	54	79
9. Wm. H. Harrison	W	Va.	1841	68	68
10. John Tyler	W	Va.	1841	51	71
11. James K. Polk	D	N.C.	1845	49	53
12. Zachary Taylor	W	Va.	1849	64	65
13. Millard Fillmore	W	N.Y.	1850	50	74
14. Franklin Pierce	D	N.H.	1853	48	64
15. James Buchanan	D	Pa.	1857	65	77
16. Abraham Lincoln	R	Ky.	1861	52	56
17. Andrew Johnson	*	N.C.	1865	56	66
18. Ulysses S. Grant	R	Ohio	1869	46	63
19. Rutherford B. Hayes	R	Ohio	1877	54	70
20. James A. Garfield	R	Ohio	1881	49	49
21. Chester A. Arthur	R	Vt.	1881	50	56
22. Grover Cleveland	D	N.J.	1885	47	71
23. Benjamin Harrison	R	Ohio	1889	55	67
24. Grover Cleveland	D	N.J.	1893	55	71
25. William McKinley	R	Ohio	1897	54	58
26. Theodore Roosevelt	R	N.Y.	1901	42	60
27. William H. Taft	R	Ohio	1909	51	72
28. Woodrow Wilson	D	Va.	1913	56	67
29. Warren G. Harding	R	Ohio	1921	55	57
30. Calvin Coolidge	R	Vt.	1923	51	60
31. Herbert C. Hoover	R	Iowa	1929	54	90
32. Franklin Roosevelt	D	N.Y.	1933	51	63
33. Harry S. Truman	D	Mo.	1945	60	88
34. Dwight Eisenhower	R	Texas	1953	62	78
35. John F. Kennedy	D	Mass.	1961	43	46
36. Lyndon B. Johnson	D	Texas	1963	55	64
37. Richard M. Nixon	R	Calif.	1969	56	
38. Gerald R. Ford	R	Mich.	1974	61	
39. Jimmy Carter	D	Ga.	1977	52	
40. Ronald Reagan	R	Ohio	1981	69	

Party designations: F—Federalist; D—Democrat; R — Republican; D-R — Democrat-Republican; W—Whig. *—Andrew Johnson, D, was nominated by Republicans, elected with Lincoln. Grover Cleveland is officially 22nd and 24th president.

A Bow to the Cow

No person should grow up without seeing a cow, watching her chew her cud, noting the way her tail swings to ward off flies, and listening to the age-old vocalizing on the note of Moo.

Cows are part of our heritage, taken for granted in every generation as basic to civilized living, but now becoming somewhat remote. Much as we hate to admit, children CAN grow up without seeing a cow; they can survive without drinking cow's milk, and they need not hear a Moo in order to develop a vocabulary of sounds.

Man believed for centuries that the milk from cows was essential to good health and growth, but attacks have been made on this premise in recent years. We are told that cow's milk may be harmful rather than helpful — yet most of us persist in belief that the cow will be vindicated.

Poets have sung about cows from time immemorial, and in this issue we gather some of the verse that the cow has inspired.

Possibly the oldest is the nursery rhyme, which may be new to some of our younger readers:

High diddle diddle
The cat and the fiddle,
The cow jumped over the moon;
The little dog laughed
To see such sport
And the dish ran away with the spoon.

In more modern times, the most famous four lines ever written about the cow were those by Gelett Burgess, penned in 1895:

I never saw a purple cow,
I never hope to see one;
But I can tell you, anyhow,
I'd rather see than be one.

This poem proved so popular, and so many persons apparently recited it to Burgess, that he wrote a few years later:

Ah, yes, I wrote the "Purple Cow" –
I'm sorry now, I wrote it!
But I can tell you, anyhow,
I'll kill you if you quote it!

A later poem, by Ogden Nash, has also become famous. Nash wrote:

The cow is of the bovine ilk,
One end is moo, the other, milk.

Cows figure prominently in proverbs. The British say that someone is "always behind, like the cow's tail," or that "a cow doesn't know what her tail is worth, 'til she loses it." When we remark something will not happen "until the cows come home," that may be a very long time indeed.

If someone is called a "sacred cow" that means it is not to be interfered with — referring to the reverence for cows in India. The expression, "Holy cow!" is a moderate form of cussing. People who like to repeat the same sound sometimes ask, "How now, brown cow?" — in other words, what's next?

Wisconsin, which prides itself as the Dairy State since it leads all others in milk and cheese production, has dubbed the dairy cow its state domestic animal. Along with its state insect, the honeybee, you can see this is a land flowing with milk and honey.

Wisconsin had more milk cows than any other state in 1978 — a total of 1,813,000 (you count the legs and divide by four). No. 1 county in the U.S. was Calumet, in Wisconsin, with 88.64 cows per square mile. Lancaster County, Pa. (our home county) was second with 86.99 per square mile.

The cow has been called "the foster mother to all mankind." The cow deserves a bow — and that's what we're offering right now!

A Prayer for All Seasons

O Lord, let me remember . . .

When the snow blots out the landscape, and the ice covers the earth like a glassy sheath, that after spring finally arrives the grass will grow green, the flowering trees will display their colorful finery, and the snowdrop, crocus, daffodil and tulip will shine again. Let me not forget, when the chilling winds turn my nose and fingers red, and fuel disappears from the burner as if it were vapor, that the seeds I plant will produce new crops, and my perennials will be revived anew. When I despair of ever seeing the sun, let me remember that it will be back with heat and warmth in just a little while. When I wrestle with the February blues, and find myself nearly unable to cope, let me remember that spring will bring renaissance to man as well as earth, and that tedium will be replaced by happy expectations of better things to come.

O Lord, let me remember . . .

When the rains fall, and the creeks run high with melting snow, bringing threats of flood and loss of topsoil and general damage, that the skies will dry and that the eternal cycle will continue to prevail with ample growth. Let me not forget, when sudden late frosts kill tender plantings, that frosts will end at last and the temperatures will ascend with every day. Let me remember, when late snows cover the earth I have plowed so carefully, that snow feeds vital nitrate into the soil, and today's mud will be tomorrow's earthen cradle of growth. When March winds rattle the windows, let me know that the noise will subside at last. When dandelions rear their yellow heads all over my greening lawn, let me recognize that dandelions can make excellent salad — and delightful wine.

O Lord, let me remember . . .

When June starts sunny and hot, and succeeding months follow its example, that these are days I yearned for all last winter, and that I can succor my plants and still find time to take a swim or lie in the shade on a quiet afternoon. When insects buzz about my head, and take bites out of my flowers and vegetables, let me welcome such friends as purple martins and ladybugs and praying mantis, all of which have hearty appetites for pests. When

Albrecht Durer

blackspot infects my roses, let me be mindful that there are remedies which can be applied. And when my back aches, and my feet grow weary, and my sweat runneth over my eyes, let me be mindful of the joy I know in seeing my treasured flowers come into beautiful bloom, and the happiness I experience in eating my own crisp lettuce, tomatoes, beans, squash, beets and other choice vegetables. When hail perforates decorative leaves, and rabbits chew off prized blooms, let me remember that if Thou art with me I will still have enough to eat fresh, preserve, store and freeze for the long cold months ahead.

O Lord, let me remember . . .

When the summer birds leave, and killing frost comes before I am ready, let me be thankful that my wood supply is ample and will warm me on cozy nights at the fireside. Let me be mindful, when winds whisk thousands of dry leaves over the yard and patio, that leaves make good compost. When snow closes the driveway and cars and school buses can't travel the roads, let me look again at the snowy whiteness and its beauty under a star-filled inky night sky. As I count up my garden profits and losses, and find that cabbage cost me a dollar a head and roses a quarter a flower, let me look forward to the joy of spading the soil again and finding it ready for planting's rites. Whenever I grow weary or discouraged, let me be mindful of my many joys. Let me ever render thanks to Thee for making Earth a garden for my neighbor and me to share.

JANUARY, 1st Month

Weeks and Days	Remarkable Days	The Moon: South morn. h.m.	The Moon: Place & Age	The Moon: rises eve. h.m.	Miscellaneous Particulars		The Sun: slo. m.	The Sun: rises h. m.	The Sun: sets h. m.
1)	*1st Sunday after Christmas*				Day's length 9 hours 23 minutes				
Sunday	1 *New Year*	10 45	♐ 28	10 45	♈ sets 2:51 mo.	☋	3	7 22	4 45
Monday	2 Abel, Seth	11 38	♐ 29	7 00	Sirius rises 6:58 ev.		4	7 22	4 46
Tuesday	3 Enoch	eve.	♑ 30	sets eve.	⊕ closest to ☉	⌣	4	7 22	4 47
Wednes.	4 Methusaleh	1 22	♑ 1	6 06	Perseus Clusters mer. 7:29 e.		4	7 22	4 48
Thursday	5 Simon	2 11	♒ 2	7 06	Bellarix mer. 10:28 ev.		5	7 22	4 48
Friday	6 *Epiphany*	2 56	♒ 3	8 06	Procyon rises 6:20 ev.		5	7 22	4 49
Saturday	7 Isador	3 40	♒ 4	9 06	Fomalhaut sets 7:58 ev.		6	7 22	4 50
2)	*1st Sunday after Epiphany*				Day's length 9 hours 29 minutes				
Sunday	8 Erhart	4 21	♓ 5	10 05	Whale's Tail sets 10:31 ev.		6	7 22	4 51
Monday	9 Julian	5 02	♓ 6	11 03	Orion Nebula mer. 10:23 ev.		7	7 22	4 52
Tuesday	10 Paul, Herm.	5 43	♈ 7	morn.	Castor mer. 12:18 m. Cl.	☌ ♀ ♅	7	7 22	4 53
Wednes.	11 Hyginus	6 24	♈ 8	12 03	♀ rises 4:34 mo.		8	7 22	4 54
Thursday	12 Rinehold	7 08	♈ 9	1 04	Southern Cross mer. 5:02 m.		8	7 21	4 55
Friday	13 Hilary	7 56	♉ 10	2 07	Arcturus rises 11:39 ev.		8	7 21	4 56
Saturday	14 Felix	8 48	♉ 11	3 13	Cor Caroli mer. 5:24 mo.		9	7 21	4 57
3)	*2nd Sunday after Epiphany*				Day's length 9 hours 39 minutes				
Sunday	15 Maurice	9 46	♊ 12	4 23	Andromeda Galaxy sets 2:15 mo.		9	7 20	4 49
Monday	16 Marcelles	10 50	♊ 13	5 32	Pleiades sets 3:35 mo.	☊	9	7 20	5 00
Tuesday	17 *Franklin born*	11 44	♋ 14	6 38	Betelguese mer. 10:11 ev.	⌢	10	7 20	5 01
Wednes.	18 Prisca	morn.	♋ 15	rises eve.	Lion's Tail rises 9:10 ev.		10	7 19	5 02
Thursday	19 *Lee born*	12 51	♌ 16	6 27	Spica rises 12:11 mo.		10	7 19	5 03
Friday	20 Sebastian	1 50	♌ 17	7 45	Adhara mer. 11:03 ev.		11	7 18	5 04
Saturday	21 Agnes	2 50	♍ 18	9 01	♂ rises 12:51 mo.		11	7 18	5 05
4)	*3rd Sunday after Epiphany*				Day's length 9 hours 49 minutes				
Sunday	22 Vincent	3 43	♍ 19	10 15	☿ Gr. Elong. 24° W. of ☉		11	7 17	5 06
Monday	23 Emerentia	4 34	♎ 20	11 25	Pollux mer. 11:37 ev.		12	7 16	5 08
Tuesday	24 *St. Catherine*	5 23	♎ 21	morn.	Regulus rises 7:15 ev.		12	7 16	5 09
Wednes.	25 *Paul's Conv.*	6 11	♏ 22	12 34	Rigel sets 2:31 mo.	☌ ♄ ☽	12	7 15	5 10
Thursday	26 Polycarpus	7 00	♏ 23	7 43	Menkalinan mer. 9:39 ev.	☌ ♀ ♃	12	7 14	5 11
Friday	27 Chrysostom	7 50	♏ 24	2 50	El Nath mer. 9:05 ev.	☌ ♅ ☽	13	7 14	5 12
Saturday	28 Charles	8 41	♐ 25	3 54	Dubhe mer. 2:36 mo.	☋	13	7 13	5 14
5)	*4th Sunday after Epiphany*				Day's length 10 hours 03 minutes				
Sunday	29 *McKinley born*	9 33	♐ 26	4 55	Vega rises 1:16 mo.	☌ ♀ ☽	13	7 12	5 15
Monday	30 Adelgunda	10 25	♑ 27	5 49	Alhena mer. 10:02 ev.	☌ ☿ ☽	13	7 11	5 16
Tuesday	31 Virgil	11 17	♑ 28	6 37	Zubenelgenubi rises 1:07 mo.	⌣	13	7 10	5 17

January has 31 days

PROBABLE STATE OF THE WEATHER

JANUARY

Synopsis: A crazy month East Coast. Everything is expected with the coldest weather in the interior; very cold and a general freeze into Florida. A very short January thaw by mid-month.

1- 3 A quick change. Rain and showers South Atlantic; snow and showers New England showers then much colder Great Plains moving south. Pacific Coast, cloudy North, clear South, then stormy.

4- 7 Wet days. Plenty of rain, fog, South Rockies across South Plains, Gulf of Mexico and Ohio Valley. Unsettled Great Lakes south and east to Coast.

8-11 Wintry time. Storm Southern Plains moves to Great Lakes; rain and snow Mississippi Valley to Great Lakes is heavy. East Coast storms also have heavy rain/snow moving into New England. Cold then clear South Rockies, North Pacific Coast east into Central Rockies.

12-15 Unsettled now. Deep storm moves northeast from Gulf of Mexico with moderate rain/snow. Another snow-bearing storm from Central Canada follows to keep North Country snowy. Colder air from Canada spreads freeze into South Texas moving east even into Florida.

16-19 More rain in East and Southwest. Another rain storm from deep South moves northeast into New England, then cold air quickly follows from Western Canada.

20-23 Rain South, clear North. Rain South Rockies and Plains; clear, cold weather in North moves east. Rain or snow Pacific Coast.

24-27 Very low temperatures. Sustained low temperatures North Plains east to Coast. Widespread fog coverage lower Ohio Valley to Gulf Coast and Central Florida before cold air spreads south.

28-31 More rain and snow. Rain and snow rather heavy Great Lakes and Ohio Valley and south. Snow threat Southern New England moves northeast.

MOON PHASES
(Eastern Standard Time)

January

New Moon the 3rd at 12 o'clock 16 minutes in the morning.

First Quarter the 11th at 4 o'clock 49 minutes in the morning.

Full Moon the 18th at 9 o'clock 6 minutes in the morning.

Last Quarter the 24th at 11 o'clock 48 minutes in the evening.

CARDINAL POINTS
Eastern Standard Time

Vernal Equinox, entrance of the Sun into Aries, March 20 at 5 o'clock 25 minutes in the morning.

Summer Solstice, entrance of the Sun into Cancer, June 21 at 12 o'clock 2 minutes in the morning.

Autumnal Equinox, entrance of the Sun into Libra, September 22 at 3 o'clock 33 minutes in the afternoon.

Winter Solstice, entrance of the Sun into Capricorn, December 21 at 11 o'clock 23 minutes in the morning.

LEAP YEAR OFFER RENEWED

A free copy of Baer's Agricultural Almanac for 1984 will be sent to every baby whose birthdate is Feb. 29, 1984. This is in keeping with a Leap Year custom of ours, and if you are in the news media we hope you will publicize it. We cannot, however, honor any birth announcement postmarked before Feb. 29 .

Leap Year, occurring each fourth year, carries with it the permission for any woman to propose marriage to the man of her choice. This tradition goes back as far as the year 1288, when Scotspersons were informed by law that in leap year each maiden lady "shal hae liberte to bespeke ye man she like, albeit he refuses to taik her to his lawfyl wyfe, he shall be mulcted in ye sum of ane pundis" (one pound, which isn't worth as much as it used to be).

FEBRUARY, 2nd Month

Weeks and Days	Remarkable Days	The Moon South eve. h.m.	Place & Age	sets eve. h.m.	Miscellaneous Particulars	The Sun slo. m.	rises h. m.	sets h. m.
Wednes.	1 Bridget	12 06	29	sets eve.	Spica rises 11:19 ev.	13	7 09	5 18
Thursday	2 *Candlemas Groundhog Day*	12 53	1	5 58	Arcturus rses 10:20 ev.	14	7 08	5 20
Friday	3 Blaisius	1 37	2	6 58	Southern Cross mer. 3:35 mo.	14	7 07	5 21
Saturday	4 Veronica	2 19	3	7 57	Capella mer. 8:21 ev.	14	7 06	5 22
6)	*5th Sunday after Epiphany*				Day's length 10 hours 18 minutes			
Sunday	5 Agatha	3 00	4	8 55	Orion Neb. mer. 8:36 ev.	14	7 05	5 23
Monday	6 Dorothy	3 40	5	9 54	Sirius sets 2:44 mo.	14	7 04	5 24
Tuesday	7 Richard	4 20	6	10 53	Menkalinan sets 6:39 mo.	14	7 03	5 26
Wednes.	8 Solomon	5 03	7	11 54	Zubeneschamali ri. 12:37 mo.	14	7 02	5 27
Thursday	9 Apolonia	5 47	8	morn.	Procyon mer. 10:24 ev.	14	7 01	5 28
Friday	10 Scholastica	6 36	9	12 58	Jupiter rises 4:23 mo.	14	7 00	5 29
Saturday	11 Euphrosina	7 29	10	2 03	Regulus mer. 12:45 mo. ☊	14	6 59	5 30
7)	*6th Sunday after Epiphany*				Day's length 10 hours 34 minutes			
Sunday	12 *Lincoln born*	8 26	11	3 10	Phecda mer. 2:27 mo.	14	6 58	5 32
Monday	13 Aurora	9 27	12	4 16	Antares rises 2:37 mo.	14	6 56	5 33
Tuesday	14 *Valentine Day*	10 33	13	5 18	Hercules Cluster mer. 7:07 mo. ☊	14	6 55	5 34
Wednes.	15 Fausinius	11 42	14	6 12	Vega rises 12:09 mo. ☌ ♂ ♄	14	6 54	5 35
Thursday	16 Julianus	morn.	15	rises eve.	Betelgeuse mer. 8:12 ev.	14	6 53	5 36
Friday	17 Constantia	12 54	16	6 34	Pollux sets 5:45 mo.	14	6 51	5 37
Saturday	18 Anselmus	1 41	17	7 52	Mirfax sets 5:05 mo.	14	6 50	5 39
8)	*7th Sunday after Epiphany*				Day's length 10 hours 51 minutes			
Sunday	19 Susannah	2 28	18	9 07	Pleiades sets 1:21 mo.	14	6 49	5 40
Monday	20 *Presidents' Day*	3 14	19	10 20	Saturn rises 11:49 ev.	14	6 47	5 41
Tuesday	21 Eleanora	4 04	20	11 31	Denebola mer. 1:47 mo.	14	6 46	5 42
Wednes.	22 *Washington b.*	4 54	21	morn.	Mergez mer. 2:09 mo. ☌ ♂ ☽	14	6 45	5 43
Thursday	23 Serenus	5 45	22	12 40	Crow mer. 2:06 mo.	14	6 43	5 44
Friday	24 Matthias	6 37	23	1 47	Sabik rises 1:51 mo. ☋	13	6 42	5 46
Saturday	25 Victor	7 30	24	2 50	Nunki rises 4:15 mo. ☌ ♆ ☽	13	6 40	5 47
9)	*8th Sunday after Epiphany*				Day's length 11 hours 09 minutes			
Sunday	26 Nestor	8 22	25	3 47	Adhara sets 12:46 mo. ☌ ♃ ☽ ☋	13	6 39	5 48
Monday	27 Leander	9 13	26	4 36	Rasalhague rises 12:26 mo.	13	6 38	5 49
Tuesday	28 Romanus	10 03	27	5 19	Albireo rises 1:15 mo.	13	6 36	5 50
Wednes.	29 *Leap Year*	10 50	28	5 55	Cor Caroli mer. 2:22 mo.	13	6 35	5 51

February has 29 days

PROBABLE STATE OF THE WEATHER

FEBRUARY

Synopsis: Record rain and snow east of Chicago to Atlantic Coast. Milder and changeable weather Rockies west; cold periods frequent Great Plains.

1- 3 Record rain and snow. Record rain and snow Ohio Valley, New England and Northern mid-Atlantic States. Heavy mountain snow Pacific Coast and Western Rockies into New Mexico. Continued very cold North Rockies to Great Lakes.

4- 7 Unsettled time. Temperatures moderate Western Great Plains to Pacific Coast. Rain and snow, very light, Rockies, local Great Lakes south. Milder air North Pacific region east over Montana.

8-11 Cold, then warm. Contrast air, cold North Rockies, warm in South creates blizard conditions east across Great Lakes into New England. Remains warm in deep South as rain spreads to Alantic Coast. Pacific Coast remains cool and wet as more cold air is poised to move south from Canada.

12-15 Settled days. Close to normal temperatures North, cold South, freeze into Florida. Cold into Great Plains, snow Great Lakes, then east. Milder Southwest into Texas.

16-19 Snow time. Unseasonable cold most of United States. Snow, ice, rain Ohio Valley into New England Coast. Floods occur near East Coast. Snow, rain, freezing rain make records New England into Tennessee.

20-23 Hot and cold. A severe coastal storm deposits heavy rain and snow New England, rain farther south. Very cold Great Plains. Clear, milder Western Rockies. Snow Ohio Valley south. Freeze in Florida, then warming.

24-29 Temperature contrast. Rain and snow Pacific and Southwestern States into Great Plains. Cloudy, then rain or snow New England, the Great Lakes to Gulf of Mexico moves to Coast. Damaging freeze into Florida.

MOON PHASES
(Eastern Standard Time)

February

New Moon the 1st at 6 o'clock 47 minutes in the evening.

First Quarter the 9th at 11 o'clock 0 minutes in the evening.

Full Moon the 16th at 7 o'clock 42 minutes in the evening.

Last Quarter the 23rd at 12 o'clock 13 minutes in the afternoon.

THE PLANETS IN 1984

Morning Stars

MERCURY — January 1 to March 8; April 22 to June 22; August 28 to October 10.

VENUS — January 1 to June 15.

MARS — May 11 to December 31.

JUPITER — June 29 to December 31.

SATURN — May 3 to November 11.

URANUS — June 1 to December 5.

NEPTUNE — June 21 to December 22.

PLUTO — April 20 to October 25.

Evening Stars

MERCURY — March 8 to April 22; June 22 to August 28; October 10 to December 31.

VENUS — June 15 to December 31.

MARS — January 1 to May 11.

JUPITER — January 1 to June 29.

SATURN — January 1 to May 3; November 11 to December 31.

URANUS — January 1 to June 1; December 5 to December 31.

NEPTUNE — January 1 to June 21; December 22 to December 31.

PLUTO — January 1 to April 20; October 25 to December 31.

(Continued on Page 21)

MARCH, 3rd Month

Weeks and Days	Remarkable Days	The Moon: South morn. h.m.	Place & Age	sets eve. h.m.	Miscellaneous Particulars	The Sun: slo. m.	rises h. m.	sets h. m.
Thursday	1 *St. David*	11 35	29	sets eve.	Venus rises 5:26 mo.	12	6 33	5 52
Friday	2 Simplicius	12 18	30	5 50	Capella sets 4:36 mo.	12	6 32	5 53
Saturday	3 Samuel	12 59	1	6 49	Arcturus rises 8:22 ev.	12	6 30	5 54
10)	*9th Sunday after Epiphany*				Day's length 11 hours 26 minutes			
Sunday	4 Adrian	1 39	2	7 47	Spica mer. 2:36 mo.	12	6 29	5 55
Monday	5 Frederick	2 20	3	8 46	Lion's Tail mer. 12:55 mo.	12	6 27	5 57
Tuesday	6 *Shrove Tues.*	3 01	4	9 47	Northern Crown rises 8:57 ev.	11	6 25	5 58
Wednes.	7 *Ash Wednes.*	3 44	5	10 49	Deneb rises 11:50 ev.	11	6 24	5 59
Thursday	8 Philemon	4 30	6	11 53	Sabik rises 12:59 mo. ☌ ☿ ☉ Sup.	11	6 22	6 00
Friday	9 Prudence	5 20	7	morn.	Dschubb rises 12:13 mo.	11	6 21	6 01
Saturday	10 Apollonius	6 14	8	12 58	South. Cross mer. 1:13 m ☊	10	6 19	6 02
11)	*1st Sunday of Lent*				Day's length 11 hours 45 minutes			
Sunday	11 Ernestus	7 12	9	2 02	♂ rises 11:14 ev.	10	6 18	6 03
Monday	12 Gregory	8 12	10	3 04	Hunting Dogs mer. 1:35 mo.	10	6 16	6 04
Tuesday	13 Emma	9 13	11	3 59	Lion's Heart sets 5:25 mo.	10	6 14	6 05
Wednes.	14 *Ember Day*	10 12	12	4 48	Alioth mer. 1:25 mo.	9	6 13	6 06
Thursday	15 Christopher	11 11	13	5 28	Zubenelgenubi rises 10:13 ev.	9	6 11	6 07
Friday	16 *Ember Day*	morn.	14	rises eve.	North. Crown mer. 3:58 mo.	9	6 10	6 08
Saturday	17 *Ember Day*	12 10	15	6 39	Ras-Algethi rises 10:44 ev.	8	6 08	6 09
12)	*2nd Sunday of Lent*				Day's length 12 hours 04 minutes			
Sunday	18 Anselmus	12 58	16	7 55	Shaula mer. 5:49 mo.	8	6 06	6 10
Monday	19 Josephus	1 50	17	9 09	Harp rises 9:59 ev.	8	6 05	6 11
Tuesday	20 *Spring begins*	2 42	18	10 23	Vernal Equinox, Spring beg. 5:25 m.	8	6 03	6 12
Wednes.	21 Benedictus	3 35	19	11 34	♃ rises 2:11 mo. ☌ ♂ ☽	7	6 01	6 13
Thursday	22 Pauline	4 28	20	morn.	Altair rises 1:20 mo. ☌ ♅ ☽ ☋	7	6 00	6 14
Friday	23 Eberhart	5 22	21	12 40	Scheat rises 3:13 mo.	7	5 58	6 15
Saturday	24 Gabriel	6 16	22	1 40	Nunki rises 2:25 mo.	6	5 57	6 16
13)	*3rd Sunday of Lent*				Day's length 12 hours 22 minutes			
Sunday	25 *Ann. V. M.*	7 09	23	2 34	Orion Neb sets 11:05 e.	6	5 55	6 17
Monday	26 Emmanuel	8 00	24	3 19	Little Dog sets 1:41 mo.	6	5 53	6 18
Tuesday	27 Gustavus	8 48	25	3 57	Northern Claw rises 9:28 ev.	5	5 52	6 20
Wednes.	28 Machus	9 33	26	4 29	Antares rises 11:43 ev.	5	5 50	6 21
Thursday	29 Guido	10 16	27	4 57	She-goat sets 2:49 mo.	5	5 48	6 22
Friday	30 Eustacius	10 58	28	5 22	Dog Star sets 11:15 ev. ☌ ♀ ☽	5	5 47	6 23
Saturday	31 Amos	11 38	29	sets eve.	Saturn rises 9:06 ev.	4	5 45	6 24

March has 31 days

PROBABLE STATE OF THE WEATHER

MARCH

Synopsis: Temperatures fluctuate rapidly. Precipitation very erratic. Much wind on Plains does damage.

1- 3 Warming days. Decreasing showers New England then warming winds west to Mississippi Valley. Storm center over Great Plains brings blizzard danger. Snow, cold Rockies. Tornadoes Texas into Ohio Valley. Rain or snow North Pacific States, then clear and cold

4- 7 Cold records remain. Another storm brings rapid changes in temperatures and rainfall over large regions from Rockies to East Coast, then another storm is near Pacific Coast by 7th.

8-11 Hot and cold. Heavy rain or snow Southern Plains, Ohio Valley into New England. Very cold air Northern Rockies continues east and south. Coast fog, rain and showers Pacific Coast. Mild and cloudiness; warm Virginia south.

12-15 Cold East, milder West. Polar air spreads to East Coast. Rain and snow cover large areas Rockies and Great Plains, milder. Tornadoes in South.

16-19 Cool Far West, warmer Deep South. Early spring storm Central Plains, then to Great Lakes to Eastern Canada deposits heavy rain, snow. Very warm air in Southeast is chilled. Temperatures near 90°F in Deep South.

20-23 Major storm in East. Growing danger of East Coast storm extending across Great Lakes. Clear cold Rockies, then coastal storm deposits rain and snow as it moves east.

24-27 Hot and cold. Very warm days Southern Plains east to Coast. Heavy rain and tornadoes Texas into Alabama, then clearing, colder with frost into Central Gulf States. Storms enter Central Pacific States, then east to Ohio Valley. Heavy rain and storms accompany them.

28-31 Heavy Midwestern storm. Snow North, heavy storms South accompany storm toward New England. Tornadoes in Texas move east, then much colder.

MOON PHASES
(Eastern Standard Time)

March

New Moon the 2nd at 1 o'clock 32 minutes in the afternoon.

First Quarter the 10th at 1 o'clock 28 minutes in the afternoon.

Full Moon the 17th at 5 o'clock 10 minutes in the morning.

Last Quarter the 24th at 3 o'clock 0 minutes in the morning.

THE PLANETS IN 1984
(Continued from Page 19)

Brightest or Best Seen

MERCURY — Low in the east before dawn in mid-February, early June, mid to late September, and late December. Low in the west in the evening twilight from mid to late March, early July, and late October.

VENUS — Unmistakable in the eastern morning sky until mid-May. More and more prominent in the western evening twilight from mid-July to year's end.

MARS — May be seen nearly all year, best in May.

JUPITER — May be seen nearly all year, best in June and July.

SATURN — May be seen from January to October, best in April and May; again in December.

Faint or Invisible

MERCURY — Except brief periods around dates above.

VENUS — May to August.

MARS — Toward year's end.

JUPITER — Toward year's end.

SATURN — October to year's end.

210
15
225

APRIL, 4th Month

Weeks and Days	Remarkable Days	The Moon: South eve. h.m.	The Moon: Place & Age	The Moon: sets eve. h.m.	Miscellaneous Particulars	The Sun: slo. m.	The Sun: rises h. m.	The Sun: sets h. m.
14)	*4th Sunday of Lent*				Day's length 12 hours 41 minutes			
Sunday	1 Theodora	12 19	♈ 30	6 40	Castor sets 3:00 mo.	4	5 44	6 25
Monday	2 Theodosia	1 00	♈ 1	7 41	☿ Gr. Elong. 19° E. of ☉	4	5 42	6 26
Tuesday	3 Ferdinand	1 43	♉ 2	8 42	Betelgeuse sets 11:32 ev.	3	5 40	6 27
Wednes.	4 Ambrosius	2 28	♉ 3	9 46	Mizar mer. 12:32 mo.	3	5 39	6 28
Thursday	5 Maximus	3 17	♉ 4	10 51	Hercules Gl. Cluster mer. 3:46 mo.	3	5 37	6 29
Friday	6 Hegisppus	4 09	♊ 5	11 55	Rasalhague mer. 4:35 m. ☊	3	5 36	6 30
Saturday	7 Aaron	5 05	♊ 6	morn.	Swan's Tail rises 9:47 ev.	2	5 34	6 31
15)	*5th Sunday of Lent*				Day's length 13 hours 00 minutes			
Sunday	8 Dionysius	6 03	♋ 7	12 57	Enif rises 2:03 mo.	2	5 32	6 32
Monday	9 Prochorus	7 01	♋ 8	1 53	Southern Fish rises 5:41 mo.	2	5 31	6 33
Tuesday	10 Daniel	7 59	♌ 9	2 42	Venus rises 4:58 mo.	1	5 29	6 34
Wednes.	11 Ezekiel	8 55	♌ 10	3 25	Eltanin mer. 4:38 mo.	1	5 28	6 35
Thursday	12 Julius	9 51	♍ 11	4 01	Southern Cross mer. 11:03 ev.	1	5 26	6 36
Friday	13 *Jefferson born*	10 46	♍ 12	4 32	Lion's Heart sets 3:22 mo.	1	5 25	6 37
Saturday	14 Tyburtis	11 30	♍ 13	5 02	Northern Cross mer. 2:04 mo.	0	5 23	6 38
16)	*Palm Sunday*				Day's length 13 hours 17 minutes			
Sunday	15 *Palm Sunday*	morn.	♎ 14	rises eve.	Vega mer. 5:02 mo.	slo	5 22	6 39
Monday	16 Calixtus	12 25	♎ 15	7 57	Dshubb ri. 9:43 ev.	fas	5 20	6 40
Tuesday	17 Rudolph	1 17	♏ 16	9 11	Arcturus mer. 12:33 mo. ☌ ♂ ☽	0	5 19	6 41
Wednes.	18 Aeneas	2 14	♏ 17	10 23	Shaula rises 12:24 mo. ☌ ♅ ☽	1	5 17	6 42
Thursday	19 *Maundy Thurs.*	3 10	♐ 18	11 28	Northern Cross mer. 5:40 mo. ☋	1	5 16	6 43
Friday	20 *Good Friday*	4 05	♐ 19	morn.	♂ rises 8:50 ev. ☌ ♆ ☽	1	5 14	6 44
Saturday	21 Adolarius	5 00	♑ 20	12 27	Antares rises 2:30 mo. ☌ ♃ ☽	1	5 13	6 45
17)	*Easter Sunday*				Day's length 13 hours 34 minutes			
Sunday	22 *Easter*	5 53	♑ 21	1 16	Lion's Tail sets 4:37 mo. ◡	1	5 12	6 46
Monday	23 *Easter Mon.*	6 43	♒ 22	1 57	Alkaid mer. 11:41 ev.	2	5 10	6 47
Tuesday	24 Albert	7 30	♒ 23	2 31	Nunki rises 12:22 mo.	2	5 09	6 48
Wednes.	25 *Mark Evan.*	8 14	♒ 24	3 01	Fomalhaut rises 4:37 mo.	2	5 07	6 49
Thursday	26 *So. Mem. Day*	8 56	♓ 25	3 27	Ras-Algethi mer. 2:56 mo.	2	5 06	6 50
Friday	27 *Grant born*	9 37	♓ 26	3 50	Swan's Head rises 9:23 ev.	2	5 05	6 51
Saturday	28 Vitalis	10 17	♈ 27	4 12	Sirius sets 9:20 ev.	2	5 03	6 52
18)	*2nd Sunday of Easter*				Day's length 13 hours 51 minutes			
Sunday	29 Sybilla	10 58	♈ 28	4 34	Pollux sets 1:01 mo.	3	5 01	6 54
Monday	30 Eutropius	11 40	♈ 29	sets eve.	♃ rises 11:43 ev.	3	5 02	6 53

Apri 13 – 57 doz.
Apr 20 – 225 doz.

Plant onions & peas –
Zuchinie – Strings Beans

April has 30 days

PROBABLE STATE OF THE WEATHER

APRIL

Synopsis: Temperatures on cool side: rain and snow.

1- 3 Cool days. Slow clearing Great Lakes. Continued snow or rain showers New England into Western North Carolina then clear and cool. Brief clear, then clouds Washington east into Dakotas, then colder. Tornadoes, rain Deep South and Georgia.

4- 7 Warm and wet. Cold air over Great Lakes moves east to New England. Storms over Southern Plains move northeast. Cold air spills from Pacific Coast across Rockies. Late rain or snow Atlantic Coast.

8-11 April showers West. Cold dry air from Northern States invades the hot moist air in South, breeding many storms. Showers and storms increase in Ohio Valley and eastward.

12-15 Heat moves north. Conflict between cold high pressure in Canada and very warm humid air in Southern Plains. Rain and storms increase in Eastern Rockies, Eastern Great Lakes into Georgia, slow to dissipate.

16-19 Strong storms, great cold. Heavy East Coast storm, another moving east from Ohio Valley. Unsettled region Pacific Coast. Very cold air moving south in mid-continent brings freeze and frost. Snow may be heavy New England to Great Lakes.

20-23 A breath of spring. Clear, warmer mid-Atlantic, Gulf of Mexico west into South Rockies, moving north. Early rains Eastern Rockies into Northern Plains move north.

24-27 Cold shocker to Great Lakes and East. Cold but dry Northern States. Rain and storms short duration Kansas and Missouri. Local storms South and West.

28-30 Tornadoes, heavy rain and snow time. Large portions of East Great Plains covered by fog, continuing east. Rocky Mountain storm brings severe weather to Plains and following cold air brings freeze from Rockies to West Plains, then east.

MOON PHASES
(Eastern Standard Time)

April

New Moon the 1st at 7 o'clock 11 minutes in the morning.

First Quarter the 8th at 11 o'clock 53 minutes in the evening.

Full Moon the 15th at 2 o'clock 11 minutes in the afternoon.

Last Quarter the 22nd at 7 o'clock 27 minutes in the evening.

New Moon the 30th at 10 o'clock 46 minutes in the evening.

CATS AND DOGS DO DREAM

Even though they can't talk about them, cats and dogs do have dreams. Scientific studies show that when these animals first go to sleep they fall into a quiet state — and then start to dream. While dreaming, dogs are more active than cats. They are probably acting out exciting or frightening experiences, such as chasing rabbits or fighting other dogs, National Geographic World magazines says. Cat may simply twitch, move their paws, or mew softly.

POPULOUS CALIFORNIA

California has been considered the most populous of the United States since 1964, when it outnumbered New York. Ratings in the 1980 census showed these states as highest: California, 23,668,562; New York, 17,557,288; Texas, 14,228,383; Pennsylvania, 11,866,728; Illinois, 11,418,461.

Plant sweet corn

MAY, 5th Month

Weeks and Days	Remarkable Days	The Moon South eve. h.m.	The Moon Place & Age	The Moon sets eve. h.m.	Miscellaneous Particulars	The Sun fast m.	The Sun rises h. m.	The Sun sets h. m.
Tuesday	1 *Philip & James*	12 25	1	7 38	Capella sets 12:39 mo.	3	5 00	6 55
Wednes.	2 Sigismund	1 13	2	8 43	Harp mer. 3:55 mo.	3	4 58	6 56
Thursday	3 *Inv. of Cross*	2 05	3	9 49	Rasalhague mer. 2:48 mo. ☊	3	4 57	6 57
Friday	4 Florianus	3 00	4	10 52	Scheat rises 12:28 mo.	3	4 56	6 58
Saturday	5 Godard	3 58	5	11 50	Whale's Tail rises 4:53 mo.	3	4 55	6 59
19)	*3rd Sunday of Easter*				Day's length 14 hours 06 minutes			
Sunday	6 Aggeus	4 56	6	morn.	rises 3:44 mo. ⌒	3	4 54	7 00
Monday	7 Domicilla	5 53	7	12 41	Demon's Eye rises 2:59 mo.	3	4 52	7 01
Tuesday	8 Stanislaus	6 49	8	1 24	Hunting Dogs mer. 9:50 ev.	4	4 51	7 02
Wednes.	9 Job	7 42	9	2 01	Crow sets 2:05 mo.	4	4 50	7 03
Thursday	10 Gorianus	8 33	10	2 33	Saturn mer. 11:29 ev.	4	4 49	7 04
Friday	11 Mamertus	9 24	11	3 02	Arcturus mer. 10:58 ev. ☍ ☿ ☉	4	4 48	7 05
Saturday	12 Pancratius	10 15	12	3 30	Zubenelgenubi mer. 11:29 ev.	4	4 47	7 06
20)	*4th Sunday of Easter*				Day's length 14 hours 21 minutes			
Sunday	13 *Mothers' Day*	11 09	13	3 58	Southern Fish rises 3:26 mo.	4	4 46	7 07
Monday	14 Christianus	11 53	14	rises eve.	Penumbral Eclipse ☌ ☿ ☽	4	4 45	7 08
Tuesday	15 Sophia	morn.	15	7 59	Dschubb sets 5:05 mo.	4	4 44	7 09
Wednes.	16 Perigrinus	12 52	16	9 10	Nunki mer. 3:18 mo. ☍ ♅ ☽ ⌣	4	4 43	7 10
Thursday	17 Jodocus	1 48	17	10 13	Altair mer. 4:09 mo. ☌ ♆ ☽	4	4 43	7 11
Friday	18 *Peace Day*	2 47	18	11 08	Alderamin me. 5:34 m. ☌ ♃ ☽ ⌣	4	4 42	7 11
Saturday	19 Portentia	3 42	19	11 53	☿ closest to ⊕ Gr. El. 26°W ☉	4	4 41	7 12
21)	*5th Sunday of Easter*				Day's length 14 hours 33 minutes			
Sunday	20 Torpetus	4 34	20	morn.	♀ rises 4:24 mo.	4	4 40	7 13
Monday	21 Hosditius	5 23	21	12 31	Eltanin mer. 1:59 mo.	3	4 39	7 14
Tuesday	22 *Nat. Maritime*	6 09	22	1 03	Spica sets 2:46 mo.	3	4 39	7 15
Wednes.	23 Desiderius	6 52	23	1 29	Northern Crown mer. 11:30 ev.	3	4 38	7 16
Thursday	24 Esther	7 33	24	1 54	Ras-Algethi mer. 1:06 mo.	3	4 37	7 17
Friday	25 Urbanus	8 13	25	2 16	Antares mer. 12:16 mo.	3	4 37	7 18
Saturday	26 Edward	8 54	26	2 38	Swan's Head mer. 3:14 m.	3	4 36	7 18
22)	*6th Sunday of Easter*				Day's length 14 hours 44 minutes			
Sunday	27 Luciaus	9 36	27	3 01	Zubeneschamali mer. 10:56 ev.	3	4 35	7 19
Monday	28 *Memorial Day*	10 19	28	3 26	Hercules Glob. Clust. mer. 12:17 mo.	3	4 35	7 20
Tuesday	29 Maxmilian	11 07	29	3 54	Fomalhaut rises 2:23 mo.	3	4 34	7 21
Wednes.	30 Beitzel	11 58	30	4 27	Annular Solar Eclipse	3	4 34	7 22
Thursday	31 Manilius	12 53	1	sets eve.	Enif rises 10:34 ev.	2	4 33	7 22

Plant second corn
Sunglow & Ister
Popcorn

May has 31 days

PROBABLE STATE OF THE WEATHER

MAY

Synopsis: Precipitation heavier East Coast, less to the west. Temperatures lower Mississippi Valley east, higher to the west.

1- 3 Stormy days in West. Rush of warm humid air north into North Dakota from Texas spawns many storms, then clear cold air follows. Stormy Great Lakes and eastward.

4- 7 Cool in Northeast and Northwest. Storm centers in New England and North Pacific prevent the central storm center from moving east. Squall line Lower Plains breeds tornadoes. Partly cloudy to clear warm South Atlantic into Mississippi and New Mexico west.

8-11 Cold to hot days. Most states east of Mississippi River are cloudy, then stormy and much colder air from Western Canada rushes across North Rockies into Great Plains to push the eastern storm to Atlantic Coast.

12-15 Milder days, cool nights. Unsettled weather, light showers, light rain by 15th near Florida and South Atlantic Coast then into New England. Unsettled Washington into Montana. Clouds increase, warmer South Pacific States east. Clear, warmer Great Lakes.

16-19 Some hot days now. Central and North Great Plains are warm, humid and stormy. Active storm center near Great Lakes breeds storms, tornadoes south to Gulf Coast. Unsettled Rockies; Pacific Coast is clear and warm.

20-23 Record cold Rockies, warm Plains and east. Tornadoes west of Mississippi River. Very warm Eastern States. Snow and freeze Rockies. Very wet South Dakota thru Oklahoma.

24-27 Hot South; cool North. Hot and humid coast to coast Deep South with local storms. Frequent rain then fresh colder air Rockies east into New England.

28-31 Showery West; slow clearing East. Cold Rockies contrast to humid Great Plains bring severe storms Oklahoma and Texas, then east. Snow Rockies. Clear, seasonal Atlantic Coast after early showers in New England.

MOON PHASES
(Eastern Standard Time)

May

First Quarter the 8th at 6 o'clock 50 minutes in the morning.

Full Moon the 14th at 11 o'clock 29 minutes in the evening.

Last Quarter the 22nd at 12 o'clock 45 minutes in the afternoon.

New Moon the 30th at 11 o'clock 49 minutes in the morning.

NO COMMENT

The editor of a small weekly newspaper in a town in the West was hard put to it one week for a matter with which to fill his columns.

So he had his composior set up the Ten Commandments, and ran them without making any editorial comment. Three days after the paper was published he received a letter saying:

"Please cancel my subscription. You're getting too personal."

COULD THIS BE YOUR AD?

LOST, yesterday, somewhere between Sunrise and Sunset, two golden hours, each set with sixty diamond minutes. No reward is offered, for they are gone forever.

Horace Mann

JUNE, 6th Month

Weeks and Days	Remarkable Days	The Moon: South eve. h.m.	Place & Age	sets eve. h.m.	Miscellaneous Particulars	The Sun: fast m.	rises h. m.	sets h. m.
Friday	1 Nicodemus	1 51	♊ 2	9 15	Vega mer. 1:57 mo. ☍ ♅ ☉ ⌓	2	4 33	7 23
Saturday	2 Marcellus	2 50	♋ 3	10 39	Northern Cross mer. 2:46 mo.	2	4 33	7 24
23)	*7th Sunday of Easter*				Day's length 14 hours 52 minutes			
Sunday	3 *Jeff Davis b.*	3 49	♋ 4	11 25	Antares mer. 11:41 ev.	2	4 32	7 24
Monday	4 Darius	4 45	♌ 5	morn.	Scheat rises 10:26 ev.	2	4 32	7 25
Tuesday	5 Boniface	5 39	♌ 6	12 03	Whale's Tail rises 2:51 mo.	2	4 32	7 26
Wednes.	6 Artensius	6 30	♍ 7	12 36	♈ rises 1:42 mo.	1	4 31	7 26
Thursday	7 Lucretia	7 19	♍ 8	1 06	Alderamin mer. 4:15 mo.	1	4 31	7 27
Friday	8 Medardus	8 08	♎ 9	1 33	Nunki rises 9:25 ev.	1	4 31	7 27
Saturday	9 Barnimus	8 58	♎ 10	2 00	♂ sets 2:27 mo.	1	4 31	7 28
24)	*Pentecost Sunday*				Day's length 14 hours 57 minutes			
Sunday	10 Flavius	9 59	♏ 11	2 28	Shaula mer. 12:17 mo. ☌ ♄ ☽	1	4 31	7 28
Monday	11 Barnabus	10 41	♏ 12	2 59	Rasalhague sets 6:58 mo.	0	4 30	7 29
Tuesday	12 Basalides	11 33	♐ 13	3 36	Altair mer. 2:27 mo. ☋	0	4 30	7 29
Wednes.	13 *Ember Day*	morn.	♐ 14	4 18	Penumbral Eclipse	fas	4 30	7 30
Thursday	14 *Flag Day*	12 33	♐ 15	rises eve.	Enif rises 4:12 mo. ☌ ♃ ☽	slo	4 30	7 30
Friday	15 *Ember Day*	1 29	♑ 16	5 46	Sabik mer. 11:35 ev. Sup. ⌣	0	4 30	7 31
Saturday	16 *Ember Day*	2 22	♑ 17	10 28	Southern Fish mer. 5:18 m.	1	4 30	7 31
25)	*Trinity Sunday*				Day's length 15 hours 00 minutes			
Sunday	17 *Fathers' Day*	3 15	♒ 18	11 02	Acturus sets 3:41 mo.	1	4 31	7 31
Monday	18 Arnolphus	4 02	♒ 19	11 30	Hercules Glob Cluster mer. 10:54 ev.	1	4 31	7 32
Tuesday	19 Gervasius	4 47	♓ 20	11 56	♃ mer. 12:48 mo.	1	4 31	7 32
Wednes.	20 Sylverius	5 29	♓ 21	morn.	Eltanin mer. 12:02 mo.	1	4 31	7 32
Thursday	21 *Summer beg.*	6 09	♓ 22	12 29	Summer begins 12:02 mo.	2	4 41	7 32
Friday	22 Achatius	6 49	♈ 23	12 41	Swan's Tail mer. 2:38 mo.	2	4 31	7 33
Saturday	23 Agrippina	7 30	♈ 24	1 03	Vega mer. 12:30 mo.	2	4 32	7 33
26)	*Corpus Christi Sunday*				Day's length 15 hours 01 minutes			
Sunday	24 *John, Bapt.*	8 12	♉ 25	1 26	Ras-Algethi mer 11:04 ev.	2	4 32	7 33
Monday	25 Elogius	8 58	♉ 26	1 53	Markab mer. 5:24 mo.	2	4 32	7 33
Tuesday	26 Jeremiah	9 47	♉ 27	2 24	Alpheratz ri. 10:13 ev. ☊	3	4 33	7 33
Wednes.	27 7-Sleepers	10 51	♊ 28	3 02	Diphda rises 1:24 m.	3	4 33	7 33
Thursday	28 Leo	11 39	♊ 29	3 48	Menkar rises 2:22 m.	3	4 33	7 33
Friday	29 *Peter & Paul*	12 39	♋ 1	sets eve.	♄ sets 1:19 mo. ☍ ♃ ☉ ⌓	3	4 34	7 33
Saturday	30 Lucina	1 39	♋ 2	9 21	Demon's Eye rises 11:27 ev.	4	4 34	7 33

June has 30 days

PROBABLE STATE OF THE WEATHER

JUNE

Synopsis: Cooler and stormy Atlantic coast; hot and dry Great Plains; near normal rain and temperatures Rockies; a bit wet and cooler Pacific Coast.

1- 3 Summer mildness. After showers and thunderstorms then clear cool Atlantic Coast. Clear cool Great Lakes and Ohio Valley. Showers and rain Texas east. Local showers cover North Pacific and Northern Rockies; unsettled South Rockies.

4- 7 Cold Rockies, much fog in East. Cold air with frost brief in Rockies. Cool Great Lakes. Rain and storms continue east thru New England south thru Georgia, then east. Storms in Florida.

8-11 Relief now, a little cooler. Clear cool air departs Great Lakes and New England, then more heat and humidity. Heavy storms with the heat from Southern Plains toward Great Lakes move east.

12-15 More heat. All Southern States coast to coast are hot and humid. Showers and thunderstorms form in the warm air as it moves north into Rockies east across Great Lakes into New England.

16-19 Sticky and wet. Widespread cloudiness Washington east thru New England. Coastal rain and fog thru Florida. Interior Atlantic Coast west thru Southern Plains, New Mexico into California hot and dry.

20-23 Hot, then cold. Clear, cool, dry days New England. More moisture Northwest Pacific. Remaining warm Southwest, South Rockies. High humidity, heavy storms and tornadoes, even fog in Gulf States.

24-27 Storms, then cool. Showers and storms Great Lakes into West Gulf States move east. Local heavy storms, tornadoes Missouri south, sweep east. Colder air North Great Plains, Rockies, West Coast, pushes to Atlantic Coast.

28-30 Real summer days. Very warm and humid most everywhere with storms most frequent Northern Plains east thru New England.

MOON PHASES
(Eastern Standard Time)

June

First Quarter the 6th at 11 o'clock 42 minutes in the morning.

Full Moon the 13th at 9 o'clock 42 minutes in the morning.

Last Quarter the 21st at 6 o'clock 11 minutes in the morning.

New Moon the 28th at 10 o'clock 20 minutes in the evening.

GARDENING
By Mary Haverstick

My father was a nursery man.
He never spoke to plants.
Mine thrive upon a lesson
Learned from him in advance.

He knew which ones like sunlight,
Which live in leafy shade,
Which prefer to shun the wind
And cower in a glade.

Sweet soil or acid, wet or dry,
He understood each need.
"A plant," he said, "in the right place
"Will flourish like a weed."

He never told them anything.
He didn't haw or hem.
I think he did a wiser thing
By listening to them.

JUMPING LIONS

Mountain lions can jump 15 to 20 feet off the ground, reports National Geographic World magazine.

YEAR WITHOUT SUMMER

On June 6, 1816, ten inches of snow fell in New England, beginning the famous "year without a summer."

JULY, 7th Month

Weeks and Days	Remarkable Days	The Moon: South eve. h.m.	The Moon: Place & Age	The Moon: sets eve. h.m.	Miscellaneous Particulars	The Sun: slo. m.	The Sun: rises h. m.	The Sun: sets h. m.
27) *3rd Sunday after Pentecost*					Day's length 14 hours 58 minutes			
Sunday	1 Theobold	2 38	3	10 04	Algenib rises 10:42 ev.	4	4 35	7 33
Monday	2 *Visit V. M.*	3 34	4	10 38	rises 2:49 mo.	4	4 35	7 33
Tuesday	3 *Dog Days beg.*	4 27	5	11 09	⊕ farthest from ⊙	4	4 36	7 32
Wednes.	4 *Independence*	5 17	6	11 37	Northern Crown sets 4:24 m.	4	4 36	7 32
Thursday	5 Demetrius	6 06	7	morn.	Northern Cr. mer. 12:36 mo.	4	4 37	7 32
Friday	6 *John Huss*	6 54	8	12 04	Antares mer. 9:31 ev.	5	4 38	7 32
Saturday	7 Edelburga	7 44	9	12 31	Enif mer. 2:42 mo. ☌ ♄ ☽	5	4 38	7 31
28) *4th Sunday after Pentecost*					Day's length 14 hours 52 minutes			
Sunday	8 Aquila	8 35	10	1 01	Markab mer. 4:33 mo.	5	4 39	7 31
Monday	9 Zeno	9 30	11	1 35	♀ rises 5:09 mo. ☌ ♅ ☽	5	4 40	7 31
Tuesday	10 Israel	10 27	12	2 14	Whale's Tail rises 12:33 mo. ☋	5	4 40	7 30
Wednes.	11 Pius	11 15	13	3 00	Crow sets 9:57 ev. ☌ ♃ ☽	5	4 41	7 30
Thursday	12 Henry	morn.	14	3 53	Crow rises 11:53 ev.	6	4 42	7 29
Friday	13 Margaret	12 12	15	rises eve.	Lion's Tail sets 11:14 e. ◡	6	4 42	7 29
Saturday	14 Bonaventure	1 02	16	9 01	Swan's Head mer. 12:01 mo.	6	4 43	7 28
29) *5th Sunday after Pentecost*					Day's length 14 hours 44 minutes			
Sunday	15 *St. Swithin*	1 54	17	9 32	Fomalhut rises 3:24 mo.	6	4 44	7 28
Monday	16 Hilary	2 41	18	9 58	Bull's Eye rises 2:00 mo.	6	4 45	7 27
Tuesday	17 Alexius	3 24	19	10 22	Andromeda Galaxy rises 7:52 ev.	6	4 45	7 26
Wednes.	18 Maternus	4 05	20	10 44	Alpheratz rises 8:46 ev.	6	4 46	7 26
Thursday	19 Rufina	4 45	21	11 06	♂ sets 11:58 ev.	6	4 47	7 25
Friday	20 Elias	5 25	22	11 28	Bellatrix rises 3:10 mo.	6	4 48	7 24
Saturday	21 Praxedes	6 06	23	11 53	Sirius rises 5:46 mo.	6	4 49	7 24
30) *6th Sunday after Pentecost*					Day's length 14 hours 33 minutes			
Sunday	22 *Mary Magd.*	6 50	24	morn.	Swan's Tail mer. 12:04 mo.	6	4 50	7 23
Monday	23 Appollinaris	7 36	25	12 21	Southern Fish rises 10:47 ev.	6	4 50	7 22
Tuesday	24 Christina	8 27	26	12 55	Scheat mer. 2:54 mo. ☊	6	4 51	7 21
Wednes.	25 *St. James*	9 23	27	1 37	Caph mer 4:00 mo.	6	4 52	7 20
Thursday	26 *St. Ann*	10 22	28	2 28	Eltanin mer. 9:40 ev.	6	4 53	7 19
Friday	27 Martha	11 23	29	3 31	Vega mer. 10:16 ev. ◠	6	4 54	7 19
Saturday	28 Pantaleon	12 24	30	sets eve.	Markab mer. 3:14 m.	6	4 55	7 18
31) *7th Sunday after Pentecost*					Day's length 14 hours 21 minutes			
Sunday	29 Beatrix	1 23	1	8 36	♃ sets 2:25 mo.	6	4 56	7 17
Monday	30 Abdon	2 19	2	9 09	Rasalhague sets 3:45 mo.	6	4 57	7 16
Tuesday	31 Germanus	3 11	3	9 39	☿ Gr. Elong 27° E. of ⊙	6	4 58	7 15

July has 31 days

PROBABLE STATE OF THE WEATHER

JULY

Synopsis: Strange weather Atlantic Coast Region and western Atlantic. Temperatures lower Northeast and Western Rockies, milder Midwest. Rainfall heavier Southeast and Lower Ohio Valley, also Southwest; drier Lower Great Plains and central Rockies.

1- 3 Clear after rain. Showers and rain New England into Florida, then clear. Clear warmer Great Plains west into Rockies. Hot, scattered storms Pacific States and Southwest.

4- 7 Storms, then clear. Many showers and storms, then changeable skies. Bright clear air quickly follows the storms.

8-11 Hot and dry South. Hot dry South and Central Plains to Gulf of Mexico. Few storms North Rockies, North Great Plains, and North Pacific States then move east. Sharp storms West Rockies, North Plains east thru New England.

12-15 Of changeable nature. Storminess at standstill, continues rains east of Great Lakes thru 14th. Heavy rains Eastern Gulf States. Hurricane off California.

16-19 Hot, then cold. Pressure centers are stronger now. Near freeze Rockies. Many rain and storm centers Great Plains east into New England. Continued warm South. Normal thunderstorms Arizona east.

20-23 Typical summer heat. Unusual heavy storms Arizona into Southern Texas, North Carolina southwest into North Florida, west into Tennessee and Kentucky.

24-27 Brief relief from heat. Heat from coast to coast south of Ohio Valley. Storms frequent Gulf Coast, Florida, Northwest Washington and Great Lakes. Hail in Michigan and Minnesota is serious. Clear, hot, then local storms New England region.

28-31 Again warmer days. Cooler drier Canadian air covers North Rockies and Northern Plains after rains. Showers and storms cover Great Lakes, the Ohio Valley east to the coast. Hot South and West. Showers and storms, then clear and real cool New England.

MOON PHASES
(Eastern Standard Time)

July

First Quarter the 5th at 4 o'clock 5 minutes in the afternoon.

Full Moon the 12th at 9 o'clock 21 minutes in the evening.

Last Quarter the 20th at 11 o'clock 2 minutes in the evening.

New Moon the 28th at 6 o'clock 52 minutes in the morning.

REMOVE FALLEN LEAVES EARLY

Leaves must be removed from lawns before they are compacted by fall rains into soggy lumps that can kill grasses in soil depressions. They must be removed promptly from newly sown grass seedings.

Greener Gardening, Easier newsletter

BEYOND DEPARTURE

"In one sense there is no death. The life of a soul on earth lasts beyond his departure. You will always feel that life touching yours, that voice speaking to you, that spirit looking out of other eyes, talking to you in the familiar things he touched, worked with, loved as familiar friends. He lives on in your life and in the lives of all others that knew him."

Angelo Patri

AUGUST, 8th Month

Weeks and Days	Remarkable Days	The Moon: South eve. h.m.	The Moon: Place & Age	The Moon: sets eve. h.m.	Miscellaneous Particulars	The Sun: slo. m.	The Sun: rises h. m.	The Sun: sets h. m.
Wednes.	1 *Lammas Day*	4 02	4	10 07	Swan's Tail mer. 12:01 mo.	6	4 58	7 14
Thursday	2 Stephen	4 51	5	10 34	Sabik sets 1:31 mo.	6	4 59	7 12
Friday	3 Augustus	5 41	6	11 03	Hercules Gl. Clust m. 7:53 e.	6	5 00	7 11
Saturday	4 Dominic	6 32	7	11 36	Nunki sets 2:24 m. ☌ ♂ ☽	6	5 01	7 10
32)	*8th Sunday after Pentecost*				Day's length 14 hours 07 minutes			
Sunday	5 Oswald	7 25	8	morn.	Southern Fish mer. 2:01 mo. ☋	6	5 02	7 09
Monday	6 *Tr. of Christ*	8 19	9	12 13	Alderamin mer. 12:19 mo.	6	5 03	7 08
Tuesday	7 Donatus	9 14	10	12 56	Enif mer. 12:40 mo. ☌ ♆ ☽	6	5 04	7 07
Wednes.	8 Emilius	10 09	11	1 47	♄ sets 10:42 ev. ◡	6	5 05	7 05
Thursday	9 Ericus	11 04	12	2 44	Antares sets 11:38 ev.	5	5 06	7 04
Friday	10 *St. Lawrence*	11 54	13	3 44	Altair mer. 10:34 ev.	5	5 07	7 03
Saturday	11 *Dog Days End*	morn.	14	4 46	Pleiades rises 10:59 ev.	5	5 08	7 02
33)	*9th Sunday after Pentecost*				Day''s length 13 hours 51 minutes			
Sunday	12 Clara	12 36	15	rises eve.	Shaula sets 11:32 ev.	5	5 09	7 00
Monday	13 Hildebert	1 18	16	8 25	Harp mer. 9:09 ev.	5	5 10	6 59
Tuesday	14 Eusebius	2 02	17	8 47	Scheat mer. 1:32 mo.	5	5 11	6 58
Wednes.	15 *Assn. V. M.*	2 43	18	9 09	Dschubb sets 11:02 ev.	4	5 12	6 56
Thursday	16 Rochus	3 22	19	9 31	Nunki mer. 9:15 ev. ☌ ♀ ☿	4	5 13	6 55
Friday	17 Bertram	4 02	20	9 53	Vega mer. 9:01 ev.	4	5 14	6 54
Saturday	18 Agepetus	4 44	21	10 20	♀ sets 7:37 ev.	4	5 15	6 52
34)	*10th Sunday after Pentecost*				Day's length 13 hours 35 minutes			
Sunday	19 *St. Joachim*	5 29	22	10 50	Alpheratz mer. 2:24 mo.	4	5 16	6 51
Monday	20 Bernard	6 17	23	11 28	Whale's Tail mer. 2:48 mo.	3	5 17	6 49
Tuesday	21 Rebecca	7 09	24	morn.	Albireo mer. 9:31 ev. ☊	3	5 17	6 48
Wednes.	22 Philibert	8 05	25	12 14	Ras-Algethi sets 2:01 mo.	3	5 18	6 46
Thursday	23 Zacchaeus	9 05	26	1 11	Alderamin mer. 11:12 ev. ◠	3	5 19	6 45
Friday	24 *St. Barhol.*	10 06	27	2 17	Markab mer. 1:28 mo.	2	5 20	6 43
Saturday	25 Ludovicus	11 06	28	3 31	Mirach mer. 2:54 mo.	2	5 21	6 42
35)	*11th Sunday after Pentecost*				Day's length 13 hours 18 minutes			
Sunday	26 Samuel	12 04	29	4 48	mer. 3:48 mo.	2	5 22	6 40
Monday	27 Gebhard	12 59	1	sets eve.	Almach mer. 3:40 m.	2	5 23	6 39
Tuesday	28 *St. Augustine*	1 52	2	8 06	♂ sets 10:19 ev. ☌ ☿ ☉ Inf.	1	5 24	6 37
Wednes.	29 *St. John beh.*	2 43	3	8 34	Bull's Eye rises 11:07 ev.	1	5 25	6 36
Thursday	30 Benjamin	3 35	4	9 03	She-goat rises 8:40 ev.	1	5 26	6 34
Friday	31 Paulinus	4 27	5	9 36	Sirius rises 3:04 mo. ☌ ♄ ☽	0	5 27	6 33

August has 31 days

PROBABLE STATE OF THE WEATHER

AUGUST

Synopsis: Watch the hurricanes this month. Hot, humid, wet periods East and South. Dry West and Plains.

1- 3 Hurricane days. Several tropical storms become hurricanes. Heavy rain on Atlantic Coast. Unsettled inland to Mississippi River. Hot humid Great Plains. Showery North Rockies to Coast. Low clouds warm South Rockies and Pacific Coast.

4- 7 A breath of autumn in North. Heavy hail and rain Eastern Plains into New Mexico. Severe hurricane South Atlantic-Eastern Gulf of Mexico moving north. Temperature extremes — cooler North, hot South.

8-11 Sluggish weather days. Fog general in South Atlantic and East Gulf States, then clearing and hot. Clear, dry California into Texas. Colder North Rockies-North Great Plains moving south behind some very active storms.

12-15 Wet in East, dry in West. Slow moving pressure center deposits bountiful rains Great Lakes south into Alabama. Dense fog in South Atlantic region. Clouds increase in New England south. Cool and dry Far West. Light local showers East Rockies and Great Plains.

16-19 Fog and sluggish heat. Only Great Lakes region has measurable rainfall. Storms are scattered west of Mississippi River. Clear and dry to Pacific Coast.

20-23 Heavy Great Lakes storms. Heavy rain moves east toward New England and Ohio Valley. Heat continues in South coast to coast. Cold air enters Rockies and Plains by 23rd.

24-27 Hot and damp, heavy rain Eastern Plains, the Ohio Valley continues to Atlantic Coast, then somewhat cooler and drier. Storms increase in South and West, then cooler.

28-31 More heat and cold. More real cold air from Canada brings more relief farther south following heavy rain and storms Great Plains, Ohio Valley, then the Coast. Hot, but dry Pacific and Southwestern States.

MOON PHASES
(Eastern Standard Time)

August

First Quarter the 3rd at 9 o'clock 34 minutes in the evening.

Full Moon the 11th at 10 o'clock 45 minutes in the morning.

Last Quarter the 19th at 2 o'clock 42 minutes in the afternoon.

New Moon the 26th at 2 o'clock 26 minutes in the afternoon.

POOR RICHARD'S START

Poor Richard's Almanac, published by Benjamin Franklin, made its first appearance at Philadelphia, Pa., on Dec. 19, 1732. Franklin continued to publish it for 25 years; copies are now quite rare and priced very high. Baer's Agricultural Almanac, a modern-day successor, costs a lot less and tries to keep you up to date, even though its wit does not compare with Ben's.

DOUBLE SIGNIFICANCE

The date of June 28 holds two special meanings in relation to World War I.

On this date in 1914, seventy years ago, occurred the event which led to the conflict — the assassination of Archduke Francis Ferdinand, of Austria, and his wife, at Sarajevo, Bosnia, by Gavrilo Princip, Serbian nationalist.

On the same date in 1919, five years later, representatives of the Allies and Germany signed the peace treaty which formally ended the war, at Versailles.

ONION STORAGE

To keep white and Bermuda onions as long as yellows, don't pack them in bags. Put them 1½" apart on screens where air can flow above and below them; keep in very dry, cool place.

Eat whites and Bermudas as soon as possible. Store yellow onions, after drying, by hanging in mesh bags in dry, airy, cool place.

Gardens for all Newsmagazine

SEPTEMBER, 9th Month

Weeks and Days	Remarkable Days	The Moon: South eve. h.m.	Place & Age	sets eve. h.m.	Miscellaneous Particulars	The Sun: slo. m.	rises h. m.	sets h. m.
Saturday	1 Aegidius	5 20	♏ 6	10 12	Orion Nebula ri. 1:10 mo.	slo	5 28	6 31
36)	*12th Sunday after Pentecost*				Day's length 13 hours 01 minutes			
Sunday	2 Hannah	6 15	♐ 7	10 54	 Menkar rises 10:02 ev. ☋	fas	5 29	6 30
Monday	3 *Labor Day*	7 10	♐ 8	11 42	Ruchbah mer. 2:34 mo.	1	5 30	6 28
Tuesday	4 Moses	8 04	♑ 9	morn.	Demon's Eye ri. 7:06 ev. ☌ ♂ ♅	1	5 31	6 26
Wednes.	5 Nathaniel	8 57	♑ 10	12 37	Bull's Eye rises 10:39 ev. ◡	1	5 32	6 25
Thursday	6 Magnus	9 47	♑ 11	1 37	Alnitak rises 12:44 mo.	2	5 33	6 23
Friday	7 Amelia	10 36	♒ 12	2 38	♃ sets 11:41 ev.	2	5 34	6 21
Saturday	8 *Nat. V. M.*	11 22	♒ 13	3 41	Vega sets 4:16 mo.	2	5 34	6 20
37)	*13th Sunday after Pentecost*				Day's length 12 hours 43 minutes			
Sunday	9 Bruno	morn.	♓ 14	4 42	She-goat rises 8:00 ev.	3	5 35	6 18
Monday	10 Pulcheria	12 02	♓ 15	rises eve.	 Procyon rises 2:03 mo.	3	5 36	6 17
Tuesday	11 Protus	12 40	♓ 16	7 13	Whale's Tail rises 8:25 ev.	3	5 37	6 15
Wednes.	12 *J. Wycliffe*	1 19	♈ 17	7 38	Hamal mer. 2:41 mo.	4	5 38	6 13
Thursday	13 Amatus	2 01	♈ 18	7 57	☿ Gt. Elong. 18° W. of ⊙	4	5 39	6 12
Friday	14 *Exalt Holy* ✠	2 42	♉ 19	8 22	Andromeda Galaxy mer. 1:08 mo.	4	5 40	6 10
Saturday	15 Nicetus	3 25	♉ 20	8 50	Altair sets 2:42 mo.	5	5 41	6 08
38)	*14th Sunday after Pentecost*				Day's length 12 hours 25 minutes			
Sunday	16 Euphemia	4 11	♉ 21	9 24	Hercules Gl. Clust. sets 1:33 m. ☊	5	5 42	6 07
Monday	17 *Citizenship D.*	5 00	♊ 22	10 05	♄ sets 8:11 ev.	5	5 43	6 05
Tuesday	18 Siegfried	5 54	♊ 23	10 56	 Deneb mer. 8:51 ev.	6	5 44	6 03
Wednes.	19 *Ember Day*	6 50	♋ 24	11 56	Nunki sets 11:23 ev. ◠	6	5 45	6 02
Thursday	20 Eustachius	7 49	♋ 25	morn.	Algenib mer. 12:15 mo.	7	5 46	6 00
Friday	21 *Ember Day*	8 48	♌ 26	1 05	Mirfak mer. 3:22 mo.	7	5 47	5 58
Saturday	22 *Ember Day*	9 56	♌ 27	2 20	Autumn begins 3:33 ev.	7	5 48	5 57
39)	*15th Sunday after Pentecost*				Day's length 12 hours 06 minutes			
Sunday	23 Hoseas	10 42	♌ 28	3 37	Pleiades rises 8:09 ev.	8	5 49	5 55
Monday	24 *St. John's Con.*	11 36	♍ 29	4 56	 Schedar mer. 12:27 mo.	8	5 50	5 53
Tuesday	25 Cleophus	12 29	♎ 1	sets eve.	Rigel rises 11:25 ev.	8	5 51	5 52
Wednes.	26 Justina	1 21	♎ 2	7 00	Pollox rises 11:37 ev. ☌ ♀ ☽	9	5 52	5 50
Thursday	27 Cosmus	2 15	♏ 3	7 32	♀ sets 6:51 ev. ☌ ♄ ☽	9	5 53	5 48
Friday	28 Wenseslaus	3 10	♏ 4	8 07	Regulus rises 2:57 mo. ☋	9	5 54	5 47
Saturday	29 *St. Michael*	4 06	♏ 5	8 49	Adhara rises 2:16 mo. ☌ ♅ ☽	10	5 54	5 45
40)	*16th Sunday after Pentecost*				Day's length 11 hours 48 minutes			
Sunday	30 Jerome	5 03	♐ 6	9 36	Betelgeuse rises 10:52 ev. ☌ ♂ ☽	10	5 55	5 43

September has 30 days

PROBABLE STATE OF THE WEATHER

SEPTEMBER

Synopsis: More hurricanes bring heavy coastal rainfall. Temperatures near seasonal levels.

1- 3 Great Lakes frost days. Record low temperatures Great Lakes, then east. Thunderstorms, some heavy Rockies and Southwest, then into Great Plains. Rain deep Southeast Atlantic States, Florida, and Gulf of Mexico Region.

4- 7 Rapid changes. Clear, colder air Great Lakes continues to East Coast after a day more of rain and showers. Continued warm in North Pacific into Arizona expands east.

8-11 Troubled waters. Both South Atlantic and Gulf of Mexico potential storm centers are moving northwest. Strong polar air spills across Great Lakes and North Plains moving south. Heavy rain east Ohio Valley, Eastern Rockies and South Plains.

12-15 A bit warmer. Cloudy Atlantic Coast. Showers and storms Great Lakes into Texas, then cooler. Clouds on Pacific Coast move inland. Clear, colder Eastern Rockies and Great Plains.

16-19 Typical Autumn weather North Pacific east into Great Plains, moving south. Squall line precedes the cooler air. Clear to partly cloudy Southwest east to Atlantic Coast.

20-23 Contrasting temperatures. Following light showers contrasting pressure centers move from North Great Plains to Atlantic Coast. Heavy rain Louisiana-Mississippi is frequent; hot, humid Texas and Florida; West Coast is hot, but dry.

24-27 Cool, then milder again. Pressure centers move from Pacific Coast east rapidly again. Rain and showers light in North, but continue heavy in South.

28-30 Big storms by 30th. Showers and rain precede sharp colder air from North and West over Pacific States and Rockies. Storm with rain and snow from Great Plains moving toward Great Lakes-Ohio Valley, then a freeze or frost into Ohio Valley, then east to New England and mid-Atlantic Coast.

MOON PHASES
(Eastern Standard Time)

September

First Quarter the 2nd at 5 o'clock 30 minutes in the morning.

Full Moon the 10th at 2 o'clock 2 minutes in the morning.

Last Quarter the 18th at 4 o'clock 32 minutes in the morning.

New Moon the 24th at 10 o'clock 12 minutes in the evening.

WASPS CAN BE ALLIES

Foraging wasps, hornets and yellow jackets hunt and kill a variety of destructive insects on garden vegetables and ornamentals. Like bees that pollinate plants, they should be appreciated as beneficial allies to the gardener, says Reed S. Roberts, extension entomologist, Utah State University.

He advises if their nest is in an out-of-the-way place, don't disturb them. He says normally wasps and bees are not aggressive and will not sting unless they or their nests are molested.

If you decide that it is necessary to control them, do so in the evening when the entire colony is in the nest and inactive. Spray directly into the nest opening with a commercial hornet and wasp killer. This should kill any that attempt to come out. Wearing protective clothing can provide added safeguard against being stung. If you see no activity around the nest for a couple of days later, you can remove the nest and destroy it.

Wasps, hornets and yellow jackets don't leave stingers when they sting. In case you do get stung while trying to remove the insects, he suggests easing the pain by applying a cold pack to the swollen area.

By contrast, the honey bee does leave its stinger along with a poison sac. If stung by one, attempt to remove the stinger without squeezing the poison sac. Scrape it off with a knife or fingernail.

The Gardener, of Men's Garden Club Inc.

OCTOBER, 10th Month

Weeks and Days	Remarkable Days	The Moon South eve. h.m.	Place & Age	sets eve. h.m.	Miscellaneous Particulars	The Sun fast m.	rises h. m.	sets h. m.
Monday	1 Remigius	5 59	7	10 30	Denebola rises 4:17 mo.	10	5 56	5 42
Tuesday	2 Vollradus	6 53	8	11 29	Elnath ri. 8:52 ev. ⌣	11	5 57	5 40
Wednes.	3 Jairus	7 45	9	morn.	Alhena rises 10:51 ev. ☌ ♂ ♆	11	5 58	5 39
Thursday	4 Franciscus	8 33	10	12 31	Diphda mer. 11:50 ev.	11	5 59	5 37
Friday	5 Placidus	9 18	11	1 33	mer. 1:10 mo.	12	6 00	5 35
Saturday	6 Fides	10 00	12	2 34	Ruchbah mer. 12:24 mo.	12	6 01	5 34
41)	*17th Sunday after Pentecost*				Day's length 11 hours 30 minutes			
Sunday	7 Amelia	10 42	13	3 34	♂ sets 9:24 ev.	12	6 02	5 32
Monday	8 *Columbus Day*	11 23	14	4 33	Aldebaran rises 8:29 ev.	12	6 03	5 31
Tuesday	9 *Fire Prev. Day*	11 56	15	5 32	Castor rises 10:15 ev.	13	6 04	5 29
Wednes.	10 Gideon	morn.	16	rises eve.	Sirius rises 12:27 mo. ☌ ☿ ☉ Sup.	13	6 06	5 28
Thursday	11 Burkhard	12 40	17	6 25	Betelguese rises 10:09 ev.	13	6 07	5 26
Friday	12 Edith	1 21	18	6 52	Alpheratz mer. 10:43 ev.	13	6 08	5 24
Saturday	13 Colomon	2 08	19	7 24	Mirfak mer. 1:55 mo. ☌ ♂ ♃	14	6 09	5 23
42)	*18th Sunday after Pentecost*				Day's length 11 hours 11 minutes			
Sunday	14 Calixtus	2 57	20	8 02	Alcyone rises 6:47 ev. ☊	14	6 10	5 21
Monday	15 Hedwig	3 48	21	8 49	Lion's Tail rises 3:22 mo.	14	6 11	5 20
Tuesday	16 Gallus	4 43	22	9 44	Crow rises 5:56 mo.	14	6 12	5 18
Wednes.	17 Florentina	5 39	23	10 48	♃ sets 9:18 ev. ⌢	15	6 13	5 17
Thursday	18 *St. Luke Ev.*	6 36	24	11 58	Bellatrix mer. 3:36 mo.	15	6 14	5 16
Friday	19 Ptolemy	7 33	25	morn.	Little Dog rises 11:29 ev.	15	6 15	5 14
Saturday	20 Felicianus	8 27	26	1 13	Alnilam rises 9:44 ev.	15	6 16	5 13
43)	*19h Sunday after Pentecost*				Day's lengh 10 hours 54 minutes			
Sunday	21 Ursula	9 20	27	2 29	Pollux rises 9:58 ev.	15	6 17	5 11
Monday	22 Cordula	10 13	28	3 45	mer. 12:03 mo.	16	6 18	5 10
Tuesday	23 Severinus	11 05	29	5 01	Caph mer. 10:01 ev.	16	6 19	5 09
Wednes.	24 *Unit. Nat. Day*	11 58	30	sets eve.	Fomalhaut mer. 8:45 ev. ☋	16	6 20	5 07
Thursday	25 Crispin	12 53	1	6 00	Menkar rises 6:33 ev. ☌ P ☉	16	6 22	5 06
Friday	26 Amandus	1 50	2	6 39	Dog Star rises 11:23 ev. ☌ ♀ ☽	16	6 23	5 05
Saturday	27 Sabina	2 48	3	7 25	♄ sets 5:45 ev. ☌ ♅ ☽	16	6 24	5 03
44)	*20th Sunday after Pentecost*				Day's length 10 hours 37 minutes			
Sunday	28 *Simon Jud.*	3 47	4	8 18	Menkalinan mer. 3:31 mo. ☌ ♆ ☽	16	6 25	5 02
Monday	29 *Zwinglius*	4 44	5	9 17	☌ ♀ ♅, ☌ ♂ ☽, ☌ ☿ ♄ ⌣	16	6 26	5 01
Tuesday	30 Serapion	5 38	6	10 19	Harp sets 12:51 mo.	16	6 27	5 00
Wednes.	31 *Hallowe'en*	6 29	7	11 23	Adhara rises 12:10 mo.	16	6 28	4 58

October has 31 days

PROBABLE STATE OF THE WEATHER

OCTOBER

Synopsis: Both temperatures and rainfall are close to normal.

1- 3 Hot, then cold. Snow and cold Rockies moving east across Northern Plains, the Great Lakes into New England. Very warm, humid air remains from Lower Rockies east across Gulf of Mexico to Atlantic Coast.

4- 7 Sunshine and rain. Atlantic Seaboard and much of Rockies are dry and cool. Many storms move east from Great Plains east to Great Lakes and Western Gulf of Mexico.

8-11 Warmer now. Substantial rains east of Great Lakes south into Alabama, then clear cooler moving east. Showers Northwest Pacific Coast are ineffective then clear, cold, all Rockies.

12-15 Elements move fast. Pressure centers move rapidly from Washington toward New England. Broad portion of country into Western Gulf of Mexico is mild with showers, then mild.

16-19 Cold East, wet Great Plains. Blustery cool northerly winds New England. Humid showery Great Lakes south. Rain or snow Western Plains into New Mexico. Clear, then showers Pacific States and Rockies.

20-23 Cool nights, mild days. Cold, northerly winds New England, then squalls. Brief cold, clear South Atlantic States and Florida. Rapid warming, then rain and storms Great Plains to Pacific Coast. Clear, hot Southwest.

24-27 Fog, then cool. Great Lakes, Ohio Valley and South Atlantic States have large fog coverage. Rain heavier east of Great Lakes, south into Florida. Texas has rain and across Central Rockies.

28-31 Cold East, milder Far West. Dry everywhere except storminess New England and North Pacific States. Watch for dust storms in North Rockies and portions of Great Plains.

SIX SUCCINCT WORDS

Brevity is the soul of wit. **William Shakespeare**

MOON PHASES
(Eastern Standard Time)

October

First Quarter the 1st at 4 o'clock 53 minutes in the afternoon.

Full Moon the 9th at 6 o'clock 59 minutes in the evening.

Last Quarter the 17th at 4 o'clock 14 minutes in the afternoon.

New Moon the 24th at 7 o'clock 9 minutes in the morning.

First Quarter the 31st at 8 o'clock 9 minutes in the morning.

ZULU PROVERBS

These Zulu proverbs were provided the Almanac by Andrew J. Torchia, head of the Associated Press Bureau in Johannesburg, South Africa:

The tree is bent while it is young.
Every leopard has a black spot
There is no river without debris.
You cannot chase two buck.
The dew cannot compete with the sun.
Do not boil the water before you grind the corn.
When a headman belches, the commoner should not notice.

EMERSON ON GARDENS

"A garden is like those pernicious machineries which catch a man's coat-shirt or his hand, and draw in his arm, his leg, and his whole body to irresistable destruction."

Ralph Waldo Emerson

NARCISSUS: FROM THE GREEK

The name of Narcissus cames from a Greek legend about a boy who so loved his own image, mirrored in a pond, that he pined away and was transformed into a flower. Some versions say he dove into the pond to be nearer to the image and drowned. Anyway, narcissus — also known as daffodil or jonquil — is today the world's favorite bulb flower.

NOVEMBER, 11th Month

Weeks and Days	Remarkable Days	The Moon: South eve. h.m.	Place & Age	sets morn. h.m.	Miscellaneous Particulars	The Sun: fast m.	rises h. m.	sets h. m.
Thursday	1 *All Saints*	7 15	8	morn.	Whale's Tail mer. 10:00 ev.	16	6 29	4 57
Friday	2 *All Souls*	7 59	9	12 26	mer. 11:19 ev.	16	6 31	4 56
Saturday	3 Hubert	8 40	10	1 26	Pleiades rises 5:28 ev.	16	6 32	4 55
45)	*21st Sunday after Pentecost*				Day's length 10 hours 21 minutes			
Sunday	4 Charlotte	9 20	11	2 25	Alpheratz sets 4:49 mo.	16	6 33	4 54
Monday	5 Malachi	9 59	12	3 24	Alhena rises 8:41 ev.	16	6 34	4 53
Tuesday	6 *Election Day*	10 40	13	4 23	♀ sets 6:42 ev.	16	6 35	4 52
Wednes.	7 Engelbert	11 25	14	5 23	She-goat mer. 2:09 mo.	16	6 36	4 51
Thursday	8 Cecelia	morn.	15	rises eve.	Penumbral Eclipse	16	6 37	4 50
Friday	9 Theodore	12 01	16	5 26	Crow rises 4:02 mo. ☊	16	6 39	4 49
Saturday	10 *Martin Luther*	12 52	17	6 02	Lion's Heart rises 12:08 m.	16	6 40	4 48
46)	*22nd Sunday after Pentecost*				Day's length 10 hours 06 minutes			
Sunday	11 *Veterans Day*	1 42	18	6 46	Almach mer. 10:41 ev. ☌ ♄ ☉	16	6 41	4 47
Monday	12 Jonas	2 39	19	7 38	Perseus Clusters mer. 10:55 ev. ⌓	16	6 42	4 46
Tuesday	13 Winebert	3 35	20	8 40	Schedar mer. 9:09 ev. ☌ ♀ ♆	16	6 43	4 45
Wednes.	14 Levinus	4 31	21	9 47	Demon's Eye mer. 11:33 e.	16	6 44	4 44
Thursday	15 Leopold	5 26	22	10 58	Bull's Eye mer. 12:57 mo. ☌ ☿ ♅	15	6 45	4 43
Friday	16 Ottomar	6 20	23	morn.	♂ sets 9:09 ev.	15	6 47	4 43
Saturday	17 Alpheus	7 12	24	12 11	Giant's Shoulder mer. 2:08 m	15	6 48	4 42
47)	*23rd Sunday after Pentecost*				Day's length 9 hours 52 minutes			
Sunday	18 Gelasius	8 02	25	1 24	Dog Star mer. 2:54 mo.	15	6 49	4 41
Monday	19 Elizabeth	8 52	26	2 38	Lion's Tail rises 1:04 mo.	15	6 50	4 40
Tuesday	20 Amos	9 43	27	3 53	Arcturus rises 3:09 mo.	14	6 51	4 40
Wednes.	21 *Off V. Mary*	10 36	28	5 08	Alkaid rises 10:32 ev.	14	6 52	4 39
Thursday	22 *Thanksgiving*	11 31	29	6 25	Total Solar Eclipse	14	6 53	4 39
Friday	23 Clement	12 30	1	sets eve.	Procyon mer. 3:29 mo. ☋	14	6 54	4 38
Saturday	24 Chrisogenes	1 29	2	6 03	Harp sets 11:12 ev. ☌ ☿ ☽	13	6 56	4 38
48)	*24th Sunday after Pentecost*				Day's length 9 hours 40 minutes			
Sunday	25 Catherine	2 29	3	7 00	☿ Gr. Elong. 22° E. of ☉	13	6 57	4 37
Monday	26 Conrad	3 26	4	8 04	♃ sets 7:12 ev. ◡	13	6 58	4 37
Tuesday	27 Jehosaphat	4 20	5	9 08	Deneb sets 2:06 mo. ☌ ♂ ☽	12	6 59	4 36
Wednes.	28 Guenther	5 09	6	10 12	Southern Fish sets 10:33 e.	12	7 00	4 36
Thursday	29 Saturninus	5 54	7	11 15	Algenib sets 2:31 mo.	12	7 01	4 36
Friday	30 *St. Andrew*	6 37	8	morn.	Aldebaran mer. 11:58 ev.	11	7 02	4 35

first snow

November has 30 days

PROBABLE STATE OF THE WEATHER

NOVEMBER

Synopsis: Seasonal in all respects but milder Western Rockies to Pacific Coast.

1- 3 It is cold in the East. Brief, but solid cold in East, especially New England into mid-Atlantic states. Rain and snow North, then unsettled storm follows from Rockies.

4- 7 Mild, then colder. Clear, crisp air mid-Atlantic and South Atlantic States then milder. Rain front from Eastern Plains moves east then clear and frosty. Rain and thunderstorms from Coastal California move east across Arizona.

8-11 Rapid changes now. Rapid temperature changes Great Lakes to Atlantic Coast. Rain or snow North is heavy, violent storms Tennessee south, all moving east to Coast. Deep freeze follows. Another storm presses inland from Washington.

12-15 It is colder now. Storm now Eastern Pacific States moves South Rockies and Great Plains toward Great Lakes and Canada, then clear deep cold into Southern States. Fog covers large areas of East. Heavy rain or snow Rockies and Great Plains.

16-19 Great Lakes storm, then very cold. Storm centers from Washington cross North Rockies give heavy snow even to Great Lakes, less across New England. Cold air follows. Late rain storms Gulf of Mexico to Florida.

20-23 Slow warming East, wet West. Conflict of very warm air from Gulf of Mexico and cold Canadian air bring rain and storms to Northern States, especially the Rockies and Great Lakes. Upper Ohio, East and Southern States are dry.

24-27 The rain comes to large areas. East of Mississippi River and Pacific States are wet, then cooler winds follow from North Rockies and Great Plains.

28-30 Watch the snow and cold. Rain and snow increase in coverage, then very cold air sweeps deeper into the South even near zero temperatures in unusual places.

MOON PHASES
(Eastern Standard Time)

November

Full Moon the 8th at 12 o'clock 43 minutes in the afternoon.

Last Quarter the 16th at 2 o'clock 0 minutes in the morning.

New Moon the 22nd at 5 o'clock 58 minutes in the afternoon.

First Quarter the 30th at 3 o'clock 1 minute in the morning.

1984 HEBREW CALENDAR

YEARS 5744-5745
The Year 5744 is the Sixth of the 303rd of 19 Years

Year	No.	Month	Day	Festival	Civil Year
5744	5	Shebat	1	Rosh-Chodesh	Thurs., Jan. 5
5744	5	Shebat	15	T'u B'Shebat	Thurs., Jan. 19
5744	6	Adar	1-2	Rosh-Chodesh ..	Fri.-Sat., Feb. 3, 4
5744	7	Adar II	1-2	Rosh-Chodesh	Sun.-Mon., Mar. 4, 5
5744	7	Adar II	13	Fast of Esther	Thurs. Mar. 15
5744	7	Adar II	14	Purim (Feast of Lots)	Sun., Mar. 18
5744	8	Nisan	1	Rosh-Chodesh	Tues., Apr. 3
5744	8	Nisan	15	First Day of Pesach (Passover)	Tues., Apr. 17
5744	9	Iyar	1-2	Rosh-Chodesh	Wed.-Thurs., May 2, 3
5744	9	Iyar	18	Lag B'Omer	Sun., May 20
5744	10	Sivan	1	Rosh-Chodesh	Fri., June 1
5744	10	Sivan	6	First Day of Shavuot (Feast of Weeks) ..	Wed., June 6
5744	11	Tammuz	1-2	Rosh-Chodesh	Sat.-Sun., June 30, July 1
5744	11	Tammuz	17	Fast of Tammuz	Tues., July 17
5744	12	Av	1	Rosh-Chodesh	Mon., July 30
5744	12	Av	9	Fast of Av	Tues., Aug. 7
5744	13	Elul	1-2	Rosh-Chodesh	Tues.-Wed., Aug. 28, 29
5745	1	Tishri	1	Rosh Hashanah (New Year)	Thurs., Sept. 27
5745	1	Tishri	3	Fast of Gedaliah	Sun., Sept. 30
5745	1	Tishri	10	Yom Kippur (Day of Atonement)	Sat., Oct. 6
5745	1	Tishri	15	First Day of Succot (Tabernacles)	Thurs., Oct. 11
5745	1	Tishri	21	Hoshanah-Rabbah ..	Wed., Oct. 17
5745	1	Tishri	22	Shemini Atzereth ..	Thurs., Oct. 18
5745	1	Tishri	23	Simchat Torah (Rejoicing of the Law)	Fri., Oct. 19
5745	2	Cheshvan	1-2	Rosh-Chodesh	Fri.-Sat., Oct. 26, 27
5745	3	Kislev	1-2	Rosh-Chodesh	Sun.-Mon., Nov. 25, 26
5745	3	Kislev	25	First Day of Chanukah (Feast of Lights)	Wed., Dec. 19
5745	4	Tebet	1-2	Rosh-Chodesh	Mon.-Tues., Dec. 24, 25

DECEMBER, 12th Month

Weeks and Days	Remarkable Days	The Moon South eve. h.m.	Place & Age	sets morn. h.m.	Miscellaneous Particulars	The Sun fast m.	rises h. m.	sets h. m.
Saturday	1 Longinus	7 17	9	12 15	Bellatrix rises 6:22 ev.	11	7 03	4 35
49)	*1st Sunday of Advent*				Day's length 9 hours 30 minutes			
Sunday	2 Candidus	7 56	10	1 14	Rigel rises 6:57 ev. ☌ ☿ ♆	11	7 04	4 35
Monday	3 Fr. Xavier	8 36	11	2 12	Alnilam rises 6:50 ev.	10	7 05	4 35
Tuesday	4 Barbara	9 18	12	3 12	Alnitak rises 6:53 ev.	10	7 06	4 35
Wednes.	5 Abigail	10 04	13	4 12	Orion Neb rises 6:56 ev. ☌ ♅ ☉	9	7 07	4 35
Thursday	6 *St. Nicholas*	10 47	14	5 15	♄ rises 5:14 mo. ☌ ☿ ♆	9	7 08	4 34
Friday	7 Ambrose	11 31	15	6 20	Betelgeuse rises 6:24 ev. ☊	9	7 09	4 34
Saturday	8 *Conc. V. M.*	morn.	16	rises eve.	Crow rises 2:07 mo.	8	7 09	4 34
50)	*2nd Sunday of Advent*				Day's length 9 hours 25 minutes			
Sunday	9 Joachim	12 32	17	5 33	Denebola rises 11:45 ev.	8	7 10	4 35
Monday	10 Judith	1 29	18	6 19	Capella mer. 11:59 ev. ⌓	7	7 11	4 35
Tuesday	11 Barsabas	2 25	19	6 51	mer. 8:46 ev.	7	7 12	4 35
Wednes.	12 Ottilia	3 22	20	8 49	Alpheratz sets 2:19 mo.	6	7 13	4 35
Thursday	13 Lucy	4 17	21	10 01	Aldebaran sets 6:04 mo.	6	7 14	4 35
Friday	14 Nicasius	5 09	22	11 31	Sirius mer. 1:12 mo.	5	7 14	4 35
Saturday	15 Ignatius	5 58	23	morn.	Procyon mer. 2:02 mo.	5	7 15	4 36
51)	*3rd Sunday of Advent*				Day's length 9 hours 20 minutes			
Sunday	16 Ananias	6 47	24	12 25	♀ sets 7:49 ev.	4	7 16	4 36
Monday	17 Lazarus	7 35	25	1 36	Castor mer. 1:49 mo.	4	7 16	4 36
Tuesday	18 Arnold	8 26	26	3 52	Arcturus rises 1:19 mo.	4	7 17	4 37
Wednes.	19 *Ember Day*	9 18	27	4 04	Regulus rises 9:34 ev. ☌ ♄ ☽	3	7 17	4 37
Thursday	20 Ammon	10 14	28	5 18	Hunting Dogs rise 10:12 ev. ☋	3	7 18	4 38
Friday	21 *Ember Day*	11 12	29	6 32	Winter begins 11:23 mo.	2	7 19	4 38
Saturday	22 *Ember Day*	12 12	30	sets eve.	Spica rises 1:58 mo.	2	7 19	4 39
52)	*4th Sunday of Advent*				Day's length 9 hours 20 minutes			
Sunday	23 Dagobert	1 11	1	5 45	Antares rises 5:59 mo. ☌ ♃ ☽	1	7 19	4 39
Monday	24 Adam, Eve	2 07	2	6 50	Pollux mer. 1:32 mo.	1	7 20	4 40
Tuesday	25 *Christmas*	2 59	3	7 56	Whale's Tail sets 11:23 ev. ☌ ♀ ☽	fas	7 20	4 40
Wednes.	26 Stephen	3 47	4	9 00	♂ sets 9:08 ev. ☌ ♂ ☽	slo	7 21	4 41
Thursday	27 *John, Evang.*	4 31	5	10 02	Algol mer. 8:43 ev.	1	7 21	4 42
Friday	28 *H. Innocents*	5 12	6	11 02	Pleiades mer. 9:19 ev.	1	7 21	4 42
Saturday	29 Noah	5 52	7	morn.	Southern Cross mer. 5:54 mo.	2	7 22	4 43
53)	*1st Sunday of Christmas*				Day's length 9 hours 22 minutes			
Sunday	30 David	6 32	8	12 01	Bellatrix rises 5:06 ev.	2	7 22	4 44
Monday	31 *N. Year's Eve*	7 12	9	12 59	Litle Dog rises 6:41 ev.	3	7 22	4 45

December has 31 days

PROBABLE STATE OF THE WEATHER

DECEMBER

Synopsis: Heavier storms Great Lakes to Gulf of Mexico, seasonal Atlantic Coast and Great Plains, but lighter to the west. Temperatures below normal coast to coast but near normal Southern California east thru South Texas east to Florida.

1- 3 Stormy time now. Western Rockies and Pacific Coast, are stormy. Rain and snow Great Lakes to Gulf Coast, then clear, cold as storm continues to the Coast.

4- 7 Heavy Gulf storm. Storm from Texas moves to Ohio Valley, then New England deposits heavy rain and snow. Very cold air follows. Freeze into Florida. Another storm from Pacific Coast moves to Eastern Rockies, then east.

8-11 More storms and cold. Rapid development of storms then quick falling temperatures affect agriculture and man.

12-15 Cold, then warmer. Great Lakes east thru New England has heavy snow prospects. Ohio Valley, Eastern Plains to Coast, it is lighter. Frigid air from Central Canada extends into Florida for two days, then becomes milder. Pacific storm brings widespread rain and snow to Rockies.

16-19 Unsettled weather now. Much cloudiness creates unsettled skies; seasonal cold nights, milder days.

20-23 A taste of winter. Waves of storminess from Alaskan waters produce heavy rain and snow squalls as a series of storms entering Washington to move east across Montana. Severe cold temperatures follow even into Florida.

24-27 Rapid weather changes. Conflict between the warm humid air from the Gulf against the storm from the Pacific Coast and the severe cold air from Canada brings stress and suffering over most of Central, Southern, and Eastern States.

28-31 Many snow squalls Northern States. Moderate rain in South, snow in North then late clear cold skies Gulf of Mexico east to Coast. Snow heavier Great Lakes and New England.

MOON PHASES
(Eastern Standard Time)

December

Full Moon the 8th at 5 o'clock 54 minutes in the morning.

Last Quarter the 15th at 10 o'clock 26 minues in the morning.

New Moon the 22nd at 6 o'clock 48 minutes in the morning.

First Quarter the 30th at 12 o'clock 28 minutes in the morning.

HALCYON DAYS

When you speak of "halcyon days," you refer to days of calm and mildness, of pleasure. That term goes back into antiquity with an origin tied to birds and weather.

The word "halcyon" is traced back to ancient Greek; it was the name of a bird, probably the kingfisher. Legend said that at the time of the winter solstice (Dec. 20-22), the halcyon fashioned its nest which floated on the sea. The bird was said to charm the waters, so that for two weeks they were placid. Today the term usually has nothing to do with birds or weather; we use it when we speak of happy times in our lives.

'GREEN SUNSHINE'

Winter's bleak landscape outside need not depress you, if you have "green sunshine" indoors in the form of house plants with bright flowers.

Some plants require minimal care, such as the self-sufficient aloe and the Christmas cactus. Others are more demanding but certainly worth the effort, such as African violets, begonias and azaleas

Poinsettias, long considered "Christmas plants" have been turned, through development of new varieties, into long-lasting houseplants not limited to one season.

Many other houseplants are available from nurserymen and florists, who will also provide suggestions for care and feeding. "Green sunshine" plants make excellent gifts.

Weather Predictions for Last Third of 1983

These forecasts for the last third of 1983 are reprinted from our previous edition for the benefit of those who made an early purchase of our 1984 edition.

SEPTEMBER

PROBABLE STATE OF THE WEATHER

1- 3 Temperatures change. A bit warmer in East. Clouds, showers eastern Plains into Great Lakes. Clearing, colder Great Plains, then frosts. Showers or storms south Plains. Unsettled eastern Rockies, showers in West to Pacific Coast.

4- 7 Storm watch. Hurricane watch deep South. Clear, cool Great Lakes, then East. Warm air, more storms Great Plains. Many clouds, warm Rockies. Storms sweep east from coast, clear later.

8-11 Fast weather changes. Pressure centers move rapidly now, so does the weather across northern States from Montana into Great Lakes. Tropical disturbances in Gulf of Mexico aproach shores between Louisiana and Texas. Fog occurs most everywhere east of Mississippi River.

12-15 Rain and storms galore. Cold and frosty Great Lakes, Ohio Valley, east to Atlantic Coast. Cooler after rain and storms South Atlantic, Florida and east Gulf of Mexico. Cloudiness, warmer, rain later Great Plains. Unsettled mid-Rockies. Very warm Southwest and Pacific Coast.

16-19 Quite warm. Light showers New England and south. Clear, cool Great Lakes west into northern Plains. Rain and storms Ohio Valley south into Florida. Rain north Pacific, also Southwest. Clear skies elsewhere.

20-23 Cool now. Clear, frosty Great Lakes south. Warmer south winds, then showery Great Plains. Tropical storm danger Florida and East Coast. Rain northwest Pacific Coast across central Rockies. Clear South and Southwest.

23-27 Tropical storms come close. South Atlantic and Gulf of Mexico very unsettled with storms that keep region wet and warm. Central Great Plains across Ohio Valley to Atlantic Coast very cool and frost damage.

28-30 Action days. Cold air from central Canada pushes freezing temperatures beyond mid-Atlantic States. Rain heavy eastern and southern States. Snow upper Rockies, cooler in South. Hard freeze northern Great Plains. Heavier showers and thunderstorms, then clear, colder central and southern Great Plains. Hard freeze Great Lakes and New England.

OCTOBER

PROBABLE STATE OF THE WEATHER

1- 3 Autumn bliss. Rain Texas, showery Great Lakes and north Great Plains. Cool Autumn winds and clearing skies Pacific Coast and central Rockies. Increasing clouds Atlantic coast.

4- 7 Many showers now. Northeast coolness is tempered as wet skies move north. Clearing skies, milder. Great Lakes, Ohio Valley and south. Sharp, frosty cold inner Mountain States of West.

8-11 Storms are brisk. Storm centers and following air move east rapidly. Clouds, milder Rockies. Coast rain and showers, partly cloudy inland. Unsettled north Great Plains, rain and showers central Plains, cool. Clear, cool after early showers New England into Virginia.

12-15 Cold, clear days. Clear, cold days are brief. Clouds become thick and frequent Great Lakes east. Showers and storms frequent Great Plains and Ohio Valley. Showery south Rockies into Gulf of Mexico region. Cold, windy, clearing north Rockies, then clear and real cold. Pacific Coast rain and stormy, gales offshore.

16-19 Wet days. Widespread rain or snow in east well into south Atlantic States. Cold freeze deep into Great Plains. Rain and storms, clear, cooler later southern Plains and lower Ohio Valley. Unsettled Great Lakes. Milder with drought-breaking rain Rockies and Pacific States.

20-23 Autumn charm time. Ocean storm Atlantic coast. Cool, few clouds east of Mississipi River. Southerly breezes, few showers, warm Great Plains. Showers increase Rockies. Late harvest rains Pacific Coast.

24-27 Cold to heat. Generous rain and showers Atlantic slope from New England to eastern Gulf states. Cold, clear air is moderated over Great Plains. Mild and cloudy Rockies. Clear, hot Southwest. Showery north, cloudy south Pacific Coast. Unsettled Great Lakes into Ohio Valley.

28-31 Sharp weather changes. The Atlantic and Pacific Coasts are cold and frosty. Eastern Great Lakes south are warmer and showery. Autumn chill, some showers and rain Great Plains. Snow threat north Rockies.

NOVEMBER

PROBABLE STATE OF THE WEATHER

1- 3 Fog and milder days. Fog in West Coast. Rain lower Mississippi Valley, eastern Lakes, and western New England. Clear, frosty Rockies. Cloudy, milder Atlantic States.

4- 7 Cooler days. Waves of showers cross northern States. Stormy Southeast and Florida. Clear, frosty Rockies. Cool and dry southern Plains and Gulf of Mexico. Showers increase north Pacific, increasing clouds elsewhere.

8-11 A bit of Indian summer. Clear, cool Pacific Coast. Cloudy most of Rockies. Local showers or rain Great Plains and Great Lakes. Increasing cloudiness, milder New England, Ohio Valley south Coastal rain, later clear, warm. Late storms in Florida.

12-15 Plenty of clouds. Heavy rain or snow north Pacific across Montana, then cool Pacific air spreads into Great Plains. Heavy rains south Atlantic and Florida subside. Cloudy, milder Great Lakes east.

16-19 Cold and stormy. Rain and storms Great Plains into Great Lakes and Ohio Valley. Clear, cold air in Rockies. Snow in northern Plains into northern New England. Clear, cool after early rain storms in Florida.

20-23 Slow recovery. A warming trend with few storms. Large areas of clear milder weather. Late stormy area northern and western Rockies.

24-27 More mildness. Warm south winds east of Mississippi River. Light showers Great Lakes and southern Plains. Clear, cold, frosty central Rockies, cloudy north and east. Fog on increase near rivers and lakes.

28-30 Cloudy; showers are widespread. Cloudy, showers north Pacific and Rockies then Great Lakes. Clear, cool east. Very warm with clouds; local storms in Great Plains move east slowly.

DECEMBER

PROBABLE STATE OF THE WEATHER

1- 3 Stormy time. Heavy snow north to central Rockies then into Great Plains. Very cold, near zero temperatures, winds north Pacific follow the rain and snow into central and southern Rockies. Increasing clouds, rain or snow New England.

4- 7 Fresh again. More rain and snow near Great Lakes, the Ohio Valley moving east. Increasing clouds, milder then rain or snow New England; rain and showers mid-Atlantic States; stormy south Atlantic states.

8-11 A bit warmer now. Clear, warmer. Clear, cold west Rockies, north Great Plains east into New England. Clear, cool lower Great Plains, Gulf of Mexico into Florida; later warmer and humid.

12-15 Wet, then cold. Rain or snow Atlantic coast before temperatures plunge. Cold and windy, clearing northern Plains and Great Lakes. Frost into Florida. Rain and storms before clearing, southern Plains and Gulf States.

16-19 Snow North, freeze South. Snow squalls Great Lakes to Atlantic Coast. Frost deep southern Plains and Gulf region. Much fog West Coast. Variable weather but cold, Rockies.

20-23 Bright days. Chinook winds north Rockies bring mildness and clouds. Only light showers North Dakota east into New England. Clear skies coast to coast in South.

24-27 Settled but cold. Light snow showers continue in north. Rain, snow, freezing rain covers a large portion east of Mississippi River. Stormy on Pacific Coast.

28-31 A deep freeze. Freezing, clear skies deep into central and eastern States. Light snow far North coast to coast. Western storm deep now in western Rockies.

Christmas Weather and Crop Proverbs

More weather proverbs seem to cluster about Christmas than any other holiday. Why? Well, Christmas is a very important observance in the year's calendar. Further, it comes just as winter is starting. when everyone is trying to peer ahead into the character of the season's weather. And, it lends itself to attempts at divination of the way crops will fare in the year to come. Here are some of the Christmas weather and crop proverbs man has devised.—Ye Ed

Warm Christmas, cold Easter; green Christmas, white Easter.

A green Christmas makes a fat graveyard.

If ice will bear a man before Christmas, it will not bear a man afterward.

The shepherd would rather see his wife enter the stable on Christmas day than the sun.

If the sun shines through the apple trees on Christmas day, there will be an abundant crop the following year.

* * *

On Christmas night, take 12 onions, naming one for each month, hollowing them slightly and filling them with salt. Put them on the windowsill overnight. In the morning, the ones in which the salt has melted will predict the wet months of the year.

* * *

Much ice on Christmas and New Year is a sign of a big fruit crop.

The length of the icicles between Christmas and New Year indicates the depth of snow during the winter.

A clear and bright sun on Christmas day fortelleth a peaceable year and plenty; but if the wind grows stormy before sunset, it betokeneth sickness in the spring and autumn quarters.

* * *

Light Christmas, light wheatsheaf (full moon near Christmas day); dark Christmas, heavy wheatsheaf.

Christmas wet, empty granary and barrel.

If it snows during Christmas night, the crops will do well.

So far as the sun shines on Christmas day, so far will the snow blow in May.

Snow on Christmas night, good hop crop next year.

If at Christmas ice hangs on the willow, clover may be cut at Easter.

When the black bird sings before Christmas, she will cry before Candlemas (Feb. 2).

Christmas goose forecast: If the breastbone of the roast goose when held up to the light shows dark upon its whole rather than otherwise, we shall have a severe winter throughout; if mottled variable, the lighter aspects betoken snow, the darker, frosts. General transparency of the bone denotes an open winter.

* * *

A windy Christmas and a calm Candlemas are signs of a good year.

If on Christmas night the wine ferments heavily in the barrels, a good wine year is to follow.

Thunder during Christmas week foretells much snow during the winter.

The 12 days between Dec. 25 and Jan. 6 are said to be the key to the whole year.

If it rains much during the 12 days after Christmas, expect a wet year.

Garden-Fresh New Mexico Cooking

Basic guidelines for cooking New Mexico style are included in the seed catalog of Plants of the Southwest, in Santa Fe. It emphasizes "quality ingredients picked fresh from the garden," and calls salsa verde (green sauce) the most important part of New Mexico cooking. The recipes are reprinted with permission from the catalog, which is filled with entrancing information. To order a catalog, send $1.00 to Plants of the Southwest, 1570 Pacheco St., Santa Fe, N.M. 87501.

SALSA VERDE

To make salsa verde, saute a small handful of fresh cilantro (coriander) or a good pinch of ground cilantro seed in oil Add a pinch of ground comino (cumin) seed, a pinch of oregano, and several cloves of minced garlic. Next, a chopped onion and about ten medium sized diced tomatillos go into the pot. It is not uncommon to throw in a potato, diced small and sauteed in oil. Stir several cups of chopped, peeled green chili. Add enough water to keep vegetables from scorching. Then dissolve about a half teaspoon of cornstarch into a half cup of cold milk. Stir this rapidly into the sauce, salt to taste, and if needed add water to give desired consistency. This sauce can be simmered until ready for use. But be forewarned: the longer it cooks, the hotter it gets!

CHEESE ENCHILADAS

The first step in making enchiladas is to fry corn tortillas (available at most supermarkets) in oil until almost crisp. Next, grate a generous portion of cheese. Monterey jack, cheddar, or meunster can be used. Chop some onion, tomatoes, and lettuce, and you are ready to assemble.

On an ovenproof plate (or in a pan from which the enchilada can be transferred onto a plate) put a tortilla, then a layer of salsa, then a layer of cheese, a sprinkling of onion, another tortilla, a layer of salsa, and some more cheese on top. Put the enchiladas in a 350° oven for several minutes until the cheese melts. Serve very hot, garnished with chopped lettuce and tomatoes. Enchiladas are sometimes served with a fried egg and/or sour cream on top. This helps cut the pungency or "hotness" of the chili.

FRIJOLES

For frijoles, soak about two cups of bolita beans in water for four hours or overnight. Mince several cloves of garlic (I always use at least 6 medium sized cloves) and saute with a little oil in a deep pot. Stir in a cup or two of diced peeled green chili, and desired seasonings (cilantro, comino, oregano, thyme, sage, or what have you). Sauteed onion also makes a good addition. Next, pour in the beans along with their soaking water. Add enough water to cover the beans an inch or two and cook over a low to medium flame until done (3-4 hours). After the first hour, add salt to taste. It takes quite a bit of salt to make any impact on a pot of beans.

Both salsa verde and frijoles can be made in advance and stored in the refrigerator for later use. They are good eaten plain, in omelettes, with hamburger, in casseroles, just about anywhere! The two most popular ways of using them are in enchiladas and burritos.

SOME FOREST FIRES HELP

Not all forest fires are bad. Some are a necessary part of the natural process that prunes and cleans the forest periodically. In some areas a lack of forest fires — the results of effective fire prevention programs — has actually inhibited the natural reseeding of certain pine trees whose seed cones need the heat of the fires to spring open, National Geographic reports.

Cheese Soup, Straws and Tart

COOK'S TOUR

Cheese Pleases

Drawing by Janice Lindstrom

Marjorie Blanchard prepared a variety of delectable cheese recipes for Country Journal magazine, and we appreciate being able to present a few of them here. She advises buying fresh cheeses in small amounts, but also suggests that if there are odd bits and pieces remaining, they can be used also in combination — in a pasta dish, an omelette or a souffle.

Edith's Cheese Straws

1 lb. sharp Cheddar cheese, grated
½ cup butter or margarine
½ teaspoon salt
¼ teaspoon cayenne or 2 drops Tabasco sauce
1¾ cups flour

Cream the cheese and butter in mixer or food processor until smooth. Beat in salt, pepper, and flour. Mix well. Roll out on floured board to ¼-inch thickness. Cut in ½- by 4-inch strips. Place on baking sheet and bake in 350° oven for 15 to 20 minutes until light brown. (These freeze well.) Yield: 100

Double Cheese Soup

1 small leek or onion
1 small carrot
1 stalk celery
3 tablespoons butter or margarine
1 tablespoon flour
1½ cups chicken or beef bouillon
1½ cups milk
Salt and freshly ground pepper to taste
1 cup grated Cheddar cheese
4 tablespoons blue cheese, crumbled

Coarsely chop vegetables. Heat butter in deep saucepan. Saute vegetables for 5 minutes. Sprinkle with flour and cook 2 minutes. Add bouillon and milk. Bring to a boil. Simmer, covered, for 20 minutes. Puree resulting soup in blender or food processor. Return to saucepan and season to taste. Stir in Cheddar cheese slowly over medium heat. Serve soup in warm bowls, adding 1 tablespoon blue cheese to each. Serves 4.

Spinach and Cheese Tart

3 packages frozen leaf spinach
4 tablespoons butter or margarine
2 shallots, minced
½ cup finely chopped onion
¼ lb. chevre cheese
3 eggs
½ lb. ricotta cheese
¼ cup cream
½ teaspon grated nutmeg
Salt and freshly ground pepper to taste

Cook spinach. Drain and dry well. Chop. Heat butter in skillet. Saute shallots and onion until soft. Combine with spinach. Crumble chevre and mix with spinach. Beat eggs into ricotta. Combine spinach and ricotta mixtures, adding enough cream to moisten and bind together. Season. Turn into buttered, shallow, 2-quart baking dish. Bake in 375° oven for 30 minutes. Serves 6.

Stuffed Endive

4 ounces Roquefort or Gorgonzola cheese
3 tablespoons buter
1 tablespoon brandy or to taste
2 heads endive

Mash together the cheese and butter with brandy and mix until smooth and creamy. Separate endive leaves, saving larger outside leaves for salad. Fill cavities of smaller leaves with cheese mixture. Serve with soup or as an hors d'oeuvre. A small can of shrimp, finely chopped, can be added to cheese mixture.

Low Cholesterol Recipes

These moderate low cholesterol, fat-controlled recipes are from the cookbook prepared by the Lipid Research Clinics of the National Institutes of Health. "Allowed oil" can be vegetable oil, corn, safflower, sunflower or soy: "allowed margarine" is corn oil margarine. The recipes are from the American Heart Association, through its chapter in Lancaster, Pa.

VEGETABLES

Vegetables can be enjoyed for their own natural flavors or they may be flavored with a variety of enhancers: Add a half teaspoon of the allowed oil to cooked, seasoned vegetables. Saute slivered almonds, mushrooms or pimento in a small amount of vegetable oil and pour over the hot vegetables. Cook vegetables in bouillon. Add touches of spices or herbs such as caraway, basil, or chives to accent flavor.

ASPARAGUS DELUXE

2 pounds fresh asparagus
¾ cup chicken bouillon (bouillon cubes may be used)
1 tablespoon cold water
½ teaspoon salt
2 tablespoons allowed oil
1 teaspoon lemon juice
1 tablespoon cornstarch

Wash asparagus and break off lower stalks. Cut diagonally in thin slices. Blend cornstarch and water and add to chicken bouillon. Cook over low heat until thickened, stirring consantly; add salt. Put oil into frying pan and heat; add asparagus slices and saute for about 3 minutes until tender. Blend lemon juice and bouillon sauce; add lo the asparagus. Stir and cook for two minutes. Serve at once. Makes 6 servings.

CARROTS DELUXE

6 medium carrots, pared in thin strips
1 tablespoon chopped parsley
1 tablespoon chopped chives
½ teaspoon salt
¼ cup corn oil margarine
1 tablespoon lemon juice
2 tablespoons water

Melt margarine in a saucepan; add lemon juice. Arrange carrots in layers in a baking dish. Between each layer pour in the lemon juice and margarine mixture and sprinkle with chives and parsley. Add water. Cover and bake at 350°F until carrots are tender, about 50 minutes. Makes 6 servings.

MARINATED VEGETABLES

Cook frozen artichoke hearts until tender. Sprinkle them with a good wine vinegar, allowed oil, coarse-ground pepper, salt and paprika. Chill for at least 1 hour. Arrange in a bowl of contrasting color, and provide colored food picks for serving. Allow 4 to 6 pieces for a serving. This is also great with cooked Brussel sprouts.

BEETS WITH HORSERADISH

2 tablespoons corn oil margarine
2 tablespoons flour
1 cup skim milk
1 medium can beets
½ teaspoon salt
Pepper to taste
Botled or grated fresh horseradish to taste (it is hot, so add with care)

Melt margarine in sauce pan but do not brown. Add the flour and when well coated in the fat, slowly add the skim milk, salt and pepper, stirring continuously. Continue sirring until thickened then add the horseradish to taste. Add beet juice and beets. Cook one minute and serve. Makes 6 servings.

BAKED SQUASH

1¼ pound yellow or green summer squash
4 teaspoons allowed oil
Salt, pepper, nutmeg
¼ cup water or bouillon

Split each squash lengthwise. Brush with oil and sprinkle with seasonings. Place in a shallow baking pan. Add water or bouillon. Cover snugly with aluminum foil. Bake in a 350°F degree oven for 30 to 35 minutes. Makes 4 servings.

Roast Goose for Winter Holidays

Roast goose is a holiday specialty that has been enjoyed by many generations of diners in the Pennsylvania Dutch country, and in Europe before them. These recipes come from Betty Groff, author of two choice volumes — "Betty Groff's Country Goodness Cookbook" and "Good Earth and Country Cooking." Betty and her husband Abram operate Groff's Farm Restaurant near Mount Joy, Pa.

ROAST GOOSE WITH POTATO AND BREAD STUFFING

1 (6-8 pounds) fresh dressed goose
Salt and pepper
1 recipe Potato and Bread Stuffing (recipe below)
2 cups water

Wash the goose and sprinkle salt and pepper inside the bird. Fill with Potato and Bread Stuffing. Liberally salt and pepper the outside. After securing the stuffed cavity with skewers, place the bird, breast side down, on a wire rack in a roasting pan. Add the water and cover, or tent with foil. Roast in a preheated 350 degree F. oven for 5 hours. Remove the cover and turn breast side up, for browning, for the last 30 minutes. Bake the extra stuffing in a buttered dish for the last 30 minutes of the baking time for the goose.

GRAVY

2 cups water
¼ cup cold water
2 tablespoonns cornstarch

Make the gravy by removing fat, adding 2 cups water to brownings and thickening with 2 tablespoons cornstarch dissolved in ¼ cup water. Stir over low heat until all the brownings have been dissolved and the mixture has thickened. Serves six.

POTATO AND BREAD STUFFING

½ cup water
½ cup chopped celery with leaves
¼ cup chopped onion
1 teaspoon salt
¼ teaspoon ground pepper
1 tablespoon chopped parsley
Pinch saffron
2 eggs, mashed potatoes
3 eggs, lightly beaten
2 cups fresh bread cubes
1 cup milk

Put the water in a 1-quart saucepan and bring to a boil. Add the celery, onion, salt, pepper, parsley and saffron. Boil approximately 7 minutes until the celery is clear. Largely mix the celery mixture with the potatoes, beaten egg, bread cubes and milk in a large bowl. If used as a side dish, bake in a preheated 350 degree F. oven for 30 minutes.

BAKED DRIED CORN PUDDING

1 cup dried corn
1½ cups milk
1 tablespoon sugar
1½ teaspons salt
Several dashes freshly ground pepper
2½ tablespoons melted butter
1½ tablespoons flour
3 eggs, beaten

Put the dried corn and milk in a quart bowl. Cover with a lid or plastic wrap and refrigerate overnight. Next day, put the sugar, salt, pepper and melted butter in a small bowl. Slowly add the flour to make a smooth paste. Add the eggs. Combine the mixture with the corn and milk. The milk should be almost completely absorbed by the corn. Pour into a buttered 1½-quart baking dish, and bake in a preheated 350 degree F. oven for 45 minutes. Serves six.

Note: For microwave, pour into the buttered 1½-quart baking dish, cover, and microwave on high for 12 minutes. Let stand 3-5 minutes before serving.

"BUY A LIVE GOOSE?"

Maryland First Settled in 1634

The year 1984 marks the 350th anniversary of the landing of the first settlers to reach Maryland, on March 25, 1634.

Arrival on St. Clements Island marked the end of a voyage that had begun 10 months before, on Nov. 22, 1623, from Cowes on the Isle of Wight. The colonists sailed two ships, the "Ark" and the "Dove."

Lord Baltimore, a distinguished nobleman who was a member of the Roman Catholic church, sought the grant of land in the New World from King James I, his friend. After James died, Baltimore continued discussions with King Charles I, but before the charter was granted, Lord Baltimore died. Charles issued the charter to Cecil, the second Lord Baltimore. Cecil's brother, Leonard Calvert, led the group of colonists, who numbered about 200.

Maryland was named in honor of Queen Henrietta Maria, who was a Catholic.

St. Mary's became the capital, but Annapolis was given this honor instead in 1699 and continues as the capital today. The city is the site of the U.S. Naval Academy. The statehouse in Annapolis is the oldest in the U.S.

Baltimore is the principal city, a major seaport, known for its renaissance centering on the harbor area and downtown business section. It was the birthplace of the Star Spangled Banner, written by Francis Scott Key in 1814 as he was held by the British, watching the bombardment of Fort McHenry.

The nation's first railroad, the Baltimore and Ohio (B&O), began in Baltimore. It has the oldest Catholic cathedral. Edgar Allan Poe and H. L. Mencken lived in Baltimore.

The succulence of its crab dishes is universally honored.

Maryland's Eastern Shore is a very special region; it qualities were highlighted in James Michener's novel, "Chesapeake."

The state's joint boundary with Pennsylvania is one of the most famous in the world — the Mason-Dixon Line.

NO WASHINGTONS IN CRYPT

Directly under the rotunda of the U.S. Capitol building in Washington is a section known as The Crypt. Under it is the vault which George and Martha Washington were to have been buried. Members of the Washington family, however, declined to give permission for removal of the bodies from Mount Vernon, and that is where they rest today

WHITE HOUSE TOURS

Do you want to visit the White House in Washington? It is open for tours Tuesday through Saturday, 10 a.m. to noon, and Saturdays, April through August, 10 a.m -2 p.m. No passes are required, but hours are subject to change without notice.

Wrens Are 'People Birds'

Of all the birds, few are so friendly to people as wrens, known for their busy nest-building and cheerful song.

Tradition says it's good luck if a wren nests near your home, and many homeowners accordingly court the wren and make it welcome.

Wrens will create a nest in any cavity — a hole in a tree trunk or a chink in a stone wall — but some are happy to find man-made boxes waiting for them, and these are the house wrens in every sense of the name.

You may think all wrens look alike, and there are only a few kinds, but Roger Tory Peterson says there are 63 species in the world. They range into North and South America. Only one, the Winter Wren, is known outside these continents — in Eurasia.

The term "jenny wren" is applied to the female, and this is an affectionate title. But there is no male counterpart — you don't hear people talking about "johnny wren."

Yet the male is very important in wren pairs. He usually arrives in the spring-summer habitat a week or so earlier than the female, and goes scouting for home sites. He may start several nests, possibly marking his territory in half a dozen bird boxes if there are that many.

The female, when she wings in, looks at the proposed nest sites and makes her choice. Several nests within a small area may look as though they are occupied, and indeed have the makings of nests, but only one really becomes the place where the eggs are laid and the young are reared.

Both adults take turns guarding the nest and feeding the new progeny. They can be observed on nearby tree limbs, one darting downward with food to enter the nest after the other emerges.

Which is Jenny? Most human observers call any wren "Jenny," and let it go at that.

The nest-building effort is the activity which gives this bird the honor of "industrious wren." Small twigs, pieces of fabric, patches of paper and string are utilized. Some birds are criticized for their shoddy nest-building; not so the wrens. They are tireless in their efforts, and wren advocates urge that 3-inch twigs be placed near nest sites to help the wrens.

Their songs vary from species to species. To many human ears, one wren's call may sound like anothers, but to the real devotees, the differences are distinct. On one wren characteristic all will agree — the wren's tail is perk-

ily erect, even if small.

They are knowledgeable about weather. Legend says that when a storm is near, the wrens stay home.

Wrens are valuable to gardeners, for they feast upon insects such as grasshoppers and beetles; they also snap up spiders, ticks and plant lice.

You don't see them at the feeders. They forage their own, enjoy keeping house whether in a thicket or a man-made box swinging from a pine limb, and sing their hearts out for the duration of their annual visits.

(We thank Dr. S. Kendrick Eshleman for his valuable counsel).

LEARNING IS LIKE A WATCH

"Wear your learning, like your watch, in a private pocket; and do not pull it out and strike it, merely to show you have one. If you are asked what o'clock it is, tell it; but do not proclaim it hourly and unasked, like the watchman."

Lord Chesterfield

CREAMERY BEGINNING

Alanson Saughter built the first "creamery," or butter factory, in the U.S. in 1861, near Wallkill, Orange County, N.Y.

SHORT MONTH

"You never realize how short a month is until you pay alimony."

John Barrymore

American Anniversaries

1534

Jacques Cartier explored the shores of the Gulf of St. Lawrence for France.

1584

Sir Walter Raleigh sent out two vessels on an exploring venture; crews sailed up the Atlantic Coast south of Chesapeake Bay and called the whole region Virginia in honor of the Virgin Queen.

1634

Ann Hutchison arrived in Boston — first clubwoman, who formed women into groups to meet for discussions at her home, and who was banished from Massachusetts because of her views.

Benjamin Syms founded first free school, in Virginia, giving 200 acres of land and "the milk of eight cows for the maintenance of an honest and learned man" to teach it.

1684

Massachusetts became a royal colony after its charter was annulled.

1734

Benjamin Franklin published "Constitution of the Freemasons," first Masonic book printed

Early locomotive, Lancaster, Pa.

Sir Walter Raleigh

in U.S.

George Boehnisch, first Moravian to reach America, arrived with a group of Schwenkfelders, Sept. 22.

Edward Blackwell wrote a book on fencing; published at Williamsburg, Va.

William and James Reese used slate for roofing material, quarried at Delta, Pa., and Cardiff, Md.

1784

John Jay named Secretary of Foreign Affairs, succeeding Robert Livingston, first to hold the position.

China trade opened; Empress of China, commanded by Capt. John Greene, reached Canton Aug. 30.

Dec. 23, New York became temporary U.S. capital.

First theological college in U.S., founded by Dutch Reformed Church, New York City.

Congress adopted decimal system of coinage.

Connecticut-Pennsylvania dispute over Wyoming Valley.

Organization of Methodist Episcopal Church in America.

About 40,000 Loyalists from U.S. settled in Nova Scotia and Upper Canada, west of Ottawa.

First French newspaper in America, Courier

de l'Amerique, published at Philadelphia, July 27.

Charles Willson Peale's Museum opened in Philadelphia, Pa.

1834

First independent law school in U.S. founded at Dickinson College, Carlisle, Pa.

McCormick reaper patented.

First medical encyclopedia published at Philadelphia.

Thomas Davenport invented street car.

Free school act passed in Pennsylvania; attempted repeal in 1835 blocked by Thaddeus Stevens.

"The Old Oaken Bucket" words written by Samuel Woodworth in 1818, set to music.

"Turkey in the Straw" composed.

First commencement held at Oberlin College, first co-ed college.

Industrial firsts — practical diving suit patented by Leonard Norcross of Dixfield, Me.; friction matches made by Daniel M. Chopin and Alonzo Dwight Phillips at Chicopee, Mass.; ice-making machine invented by Jacob Perkins, American living in London; sandpaper patent issued to Isaac Fisher, of Springfield, Vermont.

Transportation firsts — locomotive trip from Philadelphia, to Lancaster, Pa., on first state-owned railway, opened April 2, linking Philadelphia and Columbia; iron ship, "John Randolph," built at Savannah, Ga.; railroad tunnel, Hollidaysburg, Pa. to Johnstown, Pa .

Henry Blair, of Glenross, Md., became first Negro to obtain a patent, on a corn planter, Oct. 14.

Andrew Jackson was censured by Congress — for exceeding constitutional power, March 28; protest issued by Jackson, April 15.

1884

Farm prices declined after 5 years of prosperity; financial panic in New York May 5-7.

Ottmar Merganthaler invented linotype machine; revolutionized setting of type for newspapers; patented Aug. 26.

U.S. Naval War College set up at Newport, R.I.; first superintendent was Stephen Bleecker Luce, former Annapolis commandant; Oct. 6.

First New York-Boston telephone conversation.

Samuel Woodworth

Grover Cleveland (D) defeated James G. Blaine (R) for U.S. presidency.

Providence team of National League won first baseball world series title, beating Metropolitans of American Association.

Lewis E. Waterman invented fountain pen.

Congress established Bureau of Labor in the Department of the Interior.

Telephone wires first put underground.

1934

Best sellers: "The Case of the Curious Bride," Erle Stanley Gardner; "The Chinese Orange Mystery," Ellery Queen.

Rose Bowl: Columbia, 7; Stanford, 0.

World Series: St. Louis, NL, 4; Detroit, AL, 3.

Orange Bowl: Duquesne 33; Miami (Fla.) 7.

Max Baer became world heavyweight champion, knocking out Primo Carnera in 11th round, New York City, June 14.

Cavalcade won Kentucky Derby.

Nobel Award winners: Harold B. Urey, U.S., chemistry; George R. Minot, William P. Murphy, G. H. Whipple, all U.S., physiology or medicine; Luigi Pirandello, Italian, literature; Arthur Henderson, Great Britain, peace

Pulitzer Prize winners: Journalism — Medford (Ore.) Mail-Tribune, meritorious public service; Royce Brier, San Francisco Chronicle, reporting; Frederick T. Birchall, N.Y. Times, correspondence; E. P. Chase, Atlantic, Ia., News Telegraph, editorial writing; Edmund Duffy, Baltimore Sun, cartooning; Letters — Caroline Miller, "Lamb in His Bosom," fiction; Sidney Kingsley, "Men in White," drama; Herbert Agar, "The People's Choice," history; Tyler Dennett, "John Hay," biography; Robert Hillyer, "Collected Verse," American poetry.

Oscar winners: Clark Gable, actor; Claudette Colbert, actress, both in "It Happened One Night," best picture winner.

Ann Morrow Lindbergh became first woman to receive National Geographic Society gold medal as co-pilot and radio operator on Charles A. Lindbergh Aerial Survey, Jan. 30.

First high school auto driving course offered at State College High School (Pa.), under direction of Amos Earl Neyhart, Feb. 1.

Mother-in-Law Day celebrated Amarillo, Tex., March 5.

Florence Ellinwood Allen, first woman justice of U.S. Circuit Court of Appeals, April 9.

"Washeteria" — washing machines for public use — inaugurated at Fort Worth, Tex., by J. F. Cantrell, April 8.

Laurens Hammond patented pipeless organ, April 24.

"Ranger," first ship designed and built as an U.S. aircraft carrier, commissioned at Norfolk, Va. and delivered June 4.

First motion pictures of sun at Pontiac, Mich., June 19.

Rope ski tow introduced at Woodstock, Vt., June 28.

Talking Books for the Blind, New York, July.

Evangelical & Reformed Church organized, Cleveland, O., June 27; (became United Church of Christ after merger with Congregational Church, 1957).

Entire body X-rayed in 1-second exposure, Rochester, N.Y., July 1.

TWA started transcontinental overnight air transport service, Aug. 1.

First "Duck Stamp" issued, Aug. 14.

Drehersville, Pa., site of first bird sanctuary, Aug. 29.

Hawk Mountain sanctuary, Pa., founded.

Elizabeth Booth became first woman commander of Salvation Army, Sept. 3.

Morro Castle, excursion steamer, Havana-New York, burned off Asbury Park, N.J., toll 125; Sept. 8.

Nebraska became first state with a one-body legislature, Nov. 26.

National Union for Social Justice started at Royal Oak, Mich., November.

American Youth Hostels incorporated at Hartford, Conn., March 15; first hostel opened at Northfield, Mass., with Isabel and Monroe Smith as directors, Dec. 27.

Helen Richey, first woman pilot, of an airmail transport, in a Ford tri-motor, Dec. 31.

New Deal agencies formed: Export-Import Bank, Electric Farm and Home Authority, Federal Housing Administration, Federal Communications Commission, Federal Deposit Insurance Corporation, Federal Savings & Loan Insurance Corporation, U.S. Information Service, National Mediation Board, Securities & Exchange Commission.

Cyclotron developed by Prof. E. O. Lawrence, assisted by Dr. Milton Stanley Livingston, at University of California (Berkeley), to study atom's structure.

Dr. Carl David Anderson discovered the positron, a particle with a positive charge which has same mass and energy as electron, at California Institute of Technology; Anderson received Nobel Prize in Physics, 1936.

Hydroponic production of plants introduced in California.

Which Witch Is the Hazel?

Hazel trees and witch hazel plants are related only by name, yet each has witchery involved with it.

Branches of both are used in the practice of seeking underground water, which is popularly known as "dowsing," "water smelling" or "water witching." The seeker walks with a forked twig held in front of him, and when it dips (out of control), that signifies water can be found by boring straight down.

The folklore on hazel twigs for seeking water is astounding.

St. John's Eve (Dec. 26) or St. John's Night is the time to cut the twig for the purpose. But in Brandenburg, Germany, you must approach the tree in darkness, walk backward, stay silent, and cut the branch while reaching between your legs.

Berlin tradition says a good divining rod is effective only seven years, and the only one that works is one cut by an innocent child of the true faith on St. John's Night.

Long ago, hazel rods were used also to detect thieves, murderers, and treasure. If you dip the twig into water and it squeals like a pig, it is usable for further witching.

The Welsh make a "wishing cap" by weaving hazel twigs into the headgear. A sea captain with a cap like this will ride out any storm. Hazel twigs are believed to protect Irish horses and cows from fairy hi-jinks.

Witch hazel's name is attributed to the fact that it sends forth blossoms in autumn, instead of spring or summer. The American Indians, seeing this, thought the plants were bewitched. Further, the witch hazel pops its seed forth over long distances.

Witch hazel has one additional distinction: It is used for rubdown purposes. It is valued to lessen the effects of cuts and bruises. These uses are sanctioned, wise ones say, because the snakes of Aesculapius, the Roman god of the healing arts, can be seen on the flowers.

DISABILITY INSURANCE

Disability insurance can help you protect yourself against financial stress which may accompany prolonged illness or injury. About 80 per cent of all U.S. workers have some form of this type of insurance. To inform yourself, write for a booklet, "What You Should Know About Disability Insurance," to Health Insurance Association of America, 1850 K St., N.W., Washington, D.C. 20006.

Daniel Boone

DANIEL BOONE'S 250TH

Alhough Daniel Boone is associated primarily wiih Kentucky, he was born near Reading, Pa., on Nov. 2, 1734. He first saw Kentucky in 1769. The English poet Lord Byron, in his poem "Don Juan," praised Boone as "happiest among mortals anywhere," and helped bring him worldwide fame. His birthday in 1984 should be marked by civic and school observances to honor this man who typified America's pioneering spirit.

AVALANCHE!

In a typical snowy year, an estimated 100,000 avalanches occur in the United States' mountain West. The worst U.S. avalanche buried two snowbound trains near Stevens Pass in Washington in 1910, killing 96 people. The nation's worst risk area for future avalanche disasters is Alaska's capital city, Juneau, National Geographic reports.

TEDDY BEARS

Cuddled by millions of children, teddy bears were named after President Theodore Roosevelt despite the fact that he hated the nickname Teddy. The idea for the toy bear came from a 1902 political cartoon that depicted Roosevelt, the hunter, refusing to shoot a bear that had been tied to a tree for the kill, National Geographic says.

TOPS FOR ALL-AMERICA

The year 1983 marked the 50th anniversary of the All-America selections for flowers and vegetables, and four winners were named for the anniversary year.

Topping the list for 1983, each winning a bronze medal, were:

'Candlelight', ornamental pepper.
'Acapulco Silver', variegated kochia.
'Red Picotee', hybrid grandiflora petunia.
'Sweet Success', hybrid cucumber.

* * *

'Candlelight' is an edible hot pepper bearing one hundred or more slender peppers per plant. The clustered upright fruit is light green until late summer — then red in flaming color. Peppers are pencil-thin and 1½" long, sitting above the dark green foliage in majestic splendor.

Start seeds indoors at 70-75°F about 20 weeks prior to frost-free date. Expect budded plants to set outside after danger of frost. 'Candlelight' is a vigorous ornamental, resistant to sun and heat, producing fruit right up to frost.

This hardy pepper can be used en masse or as a border. The 10-12 inch plant is low and spreading, yet performs well in containers.

* * *

'Acapulco Silver' lives up to its name; its foliage is silver-tipped as though lightly hand-painted. Come fall, the bicolor silver-on-green turns maroon, then red, showing why it bears the common name of Mexican Firebush.

For best growth, sow seeds in the garden where plants will remain. If you plan to transplant, do so when seedlings are quite small with a sizeable ball of soil around the taproots. Full sun makes kochia happy.

'Acapulco Silver' grows quickly to a height of 3 or 4 feet, with width about 18 inches.

* * *

'Red Picotee' flowers break up into strongly marked blossoms with red centers and pure white white margins. They make a break from the patterned petunias of the past which reverted to solid reds and whites, with no two patterns alike.

Petunia, Grandiflora,
RED PICOTEE hybrid

Petunias are the top bedding plant flower and 'Red Picotee' is the most outstanding grandiflora of recent years. It is vigorous and easy to grow. Growers will be excited at the colorful and uniform appearance of this petunia in packs, pots, and hanging baskets.

* * *

'Sweet Success' combines so many merits in one plant that it could have been given no other name. Now delicious slicing cucumbers are no longer a luxury. The long (12-24 inches) fruit are so tender and sweet you don't have to peel them. They are "burpless," with crispy white interiors, perfect for summer salads.

'Sweet Success' has genetic resistance to cucumber mosaic virus, watermelon mosaic virus, scab, and target leaf spot, as well as tolerance to powdery mildew and downy mildew.

* * *

Look for the seeds in nurseries, garden shops and other places which specialize in the best for both new and experienced home rakers and hoers. Winners must pass rigid tests to qualify for the medals, and you and your family (and friends, and neighbors) can benefit as you harvest the crops.

STARS AS FORECASTERS

When the stars begin to huddle/the earth will soon become a puddle.

In western Kansas it is said that when the moon is near full it never storms.

When the sky seems very full of stars, expect rain, or in winter, frost.

February Lore

FLOWER FACTS & FANCIES

Julius Caesar became bald at about age 30 and was sensitive about it — so he always wore a crown of flowers or leaves to conceal his pate.

* * *

The goldenrod was the state flower of Alabama until camellia lovers — who are many in number — persuaded the Legislature to give their owner the designation in 1959.

* * *

Nero's favorite flower was the rose. Legend says he liked to wear roses, sleep on them, eat rose-flavored food, drink rose wine, and take medicine made from roses.

* * *

Evergreen was hung in Teutonic houses so that the sylvan spirits might be sheltered from winter blizzards and icy weather.

* * *

English tradition says that foxglove bells were rung by foxes when they fled from hounds. Their tinkling music bewildered the hounds and warned other animals of danger.

* * *

In ancient Greece wedding ceremonies, the priest crowned both bride and groom with wreaths of lilies and ears of corn as symbols of purity and abundance. After the wedding, the entire bridal party was crowned with flowers.

* * *

The original name of the carnation was "coronation," because it was fashioned into crowns. garlands and wreaths, and used in the crowning of kings and queens.

* * *

The botanical name of forget-me-not is Mycsotis, meaning "mouse ear" — which refers to the shape of its leaves.

* * *

The flowering almond, although closely related to the true almond, is really a cherry or plum; it does not produce fruit or nuts.

* * *

Michel Begon's name is carried on by the begonia. Begon, a French naval officer, brought the begonia from its tropical home and introduced it to cultivation.

* * *

In Maine, some folks believe that if you point at a daffodil, it won't bloom.

Camellia

Old Tennessee saying: If you chew dogwood you will lose your lover.

* * *

If you smell a lily, especially a tiger lily, you will get freckles.

* * *

Meadowsweet flowers, steeped in wine, "make the heart merrie and joyful and delight the senses," according to Gerard, medieval herbalist.

* * *

Tansy tea, made from blossoms or leaves, has been used as a tonic, or to cure stomach complaints or bathe ulcers, and even to tighten the teeth.

EARLY BLOOMER

Skunk cabbage is one of the first North American plants to bloom in the spring. Flowers inside each of the plant's red hoods burn stored energy and produce heat that melts the snow around them, National Geographic World magazine reports.

FIRE AS A FORECASTER

A fire hard to kindle indicates bad weather.

Fire is said to burn brighter and throw out more heat just before a storm.

Blacksmiths select a stormy day in which to prepare work requiring extra heat.

SUCCESS WITH GRAPES

Not everyone can grow grapes, and not everyone may want to, after reading a list of basic requirements for success prepared by the U.S. Department of Agriculture.

Here's the list, compiled by J R. McGrew, a USDA research plant pathologist at Beltsville, Md.:

- A growing season of at least 140 frost-free days.
- A site with full sunshine and good air drainage (not frosty).
- Soils that are neither waterlogged nor shallow, at least 3 feet deep.
- Willingness to spray at least 3 times a year to control insects and diseases.
- Patience to wait three to four years for vines to reach maturity before cropping.
- Annual pruning of vines
- Readiness to defend the fruit against birds by netting the vines or bagging clusters.

— If you can pass this test, be ready for one of nature's most rewarding crops — to be used fresh or stored as table fruit, to be made into jellies or juice — or fermented into wine with your very own label on it. Once you've started, all can go well.

One more hint — grow a type suitable to your area and climate, and fitted for the use to which you'll put the crop.

WORK RULES, OLD STYLE

These rules were posted in a New York City carriage shop in 1878:

Working hours shall be 7:00 a.m to 8:00 p.m. every evening but the Sabbath. On the Sabbath, everyone is expected to be in the Lord's House.

It is expected that each employee shall participate in the activities of the church and contribute liberally to the Lord's work.

All employees must show themselves worthy of their hire.

All employees are expected to be in bed by 10:00 p.m. Except: Each male employee may be given one evening a week for courting purposes and two evenings a week in the Lord's House.

It is the bounden duty of each employee to put away at least 10% of his wages for his declining years, so that he will not become a burden upon the charity of his betters.

Grapes

GET LANDSCAPING ADVICE

For those who wish to have their grounds landscaped, the American Association of Nurserymen advises five major points to bear in mind:

Beauty, comfort, convenience and safety, ease of maintenance, and sight-screening.

The time to seek advice is **before** you start the job. Many of us have no idea of the way various trees and shrubs will grow in years to come — or the way they can cancel out each other's role in the total scene. Nurserymen say the cost of aid is nominal, especially considering the harmful effects you can avert.

PLEIADES AS OMENS

If the Pleiades rise fine they set rainy; if they rise wet they set fine.

Swahili proverb

TIDES AS FORECASTERS

Showers occur more frequently at the turn of the tide.

If, during the absence of wind, the surface of the sea becomes agitated by a long rolling swell, a gale may be expected.

If, after the first ebb of the tide, it flows again for a little while, a storm approaches (Scotland).

Sylvester Graham

HOW WE GOT GRAHAM BREAD

Generations of diners have dutifully munched their graham whole wheat bread blissfully ignorant of the manner in which this staple of many diets came to be.

The bread takes its name from Sylvester Graham, born in 1794 in West Suffield, Conn. Graham, a Presbyterian minister and an advocate of food reform, spoke for vegetables as a cure for intemperance. He urged use of a flour made of whole wheat coarsely ground, for bread, muffins and rolls. His principles became known as Grahamism, and his followers were called Grahamites.

The flour he favored became known as graham flour.

He didn't mind if people laughed at him. According to the Encyclopedia Americana, he lectured in the East, favoring cold showers, hard mattresses, and loose clothing as well as unsifted flour. His subjects included cholera, comparative anatomy, personal hygiene, chastity and Biblical topics.

STUMBLE MESSAGES

If in going visiting you stumble with the right foot you will be welcome; but if you stumble with the left foot, you had better return home at once.

Pennsylvania Dutch advice

RECOVERY FOR DAMAGE

When trees or other landscape plants are damaged or destroyed, by vandalism, accident or weather calamity, you may be able to regain a significant portion of the value through your insurance policy or federal income tax consideration.

Generally, the damage must have been caused by an identifiable event of sudden, unexpected or unusual nature. Typical causes might be the chopping of an evergreen by a thief who wants it for a Christmas tree; or the smashing of a truck into a new hedge, or the effects of wind or ice storms, lightning or floods.

Your insurance agent will tell you what the policy says as to coverage; it's all there in the fine print. You may not be covered; your deductible may be higher than your dollar cost.

If you wish to seek income tax credit, make certain the damage or loss if well documented. Photos immediately after the crisis are very valuable.

An appraisal by a competent tree, nursery or landscape professional should be obtained. The professional can also advise on action to be taken, if any.

For further information, send a self-addressed, business size envelope to Tree Casualty, 232 Southern Bldg., Washington, D.C. 20005

NEWS FROM THE HENHOUSE

Mother hens and their chicks "talk" to each other even before the eggs are hatched. Researchers who placed microphons in nests have recorded peeps coming from inside eggs and clucks from responding hens. They believe these "conversations" may help a chick become familiar with its mother's voice and more responsive to her, National Geographic World magazine reports.

FLOWERS AS FORECASTERS

The perfume of flowers is more apparent just before a shower (when the air is moist) than at any other time.

If the African marigold does not open its pedals by 7 in the morning, it will rain or thunder that day.

When cowslip stalks are short, expect a dry summer.

May Lore

FLOWER NAMES VARY

What may be a cosmos to one person is a Mexican aster to another. Below are two columns with common names. Try to match them. Answers are upside down below.

A	B
Bachelor's button	Firecracker plant
Busy Lizzie	Pansy
China pink	Scotch broom
Cigar flower	Sweet William
Viola	Corn flower
Joseph's coat	Statice
Lazy daisy	Salvia
Sage	Impatiens
Sea lavender	Prairie daisy
Butcher's broom	Love-lies-bleeding

Bachelor's button—cornflower; Busy Lizzy—impatiens; China pink—sweet William; cigar flower—firecracker plant; viola—pansy; Joseph's coat—love-lies-bleeding; lazy daisy—prairie daisy; sage—salvia; sea lavender—statice; butcher's broom—scotch broom.

JASMINE

Jasmine species — fragrant blossoms make these erect or climbing shrubs a delight to the herb and spice gardener. Most have small, shiny dark green leaves. Flowers in yellow or white appear in clusters at axils or tips.

Put them in the sun and keep the soil lightly moist. Varieties of J. sambac like warmth and are good indoor plants. Propagate by cuttings of nearly mature wood.

"The World of Herbs & Spices," an Ortho Book

STEAM & GAS GUIDE

Collectors of steam traction engines, gas tractors, and stationary gas engines will want to know about the Steam & Gas Show Directory which lists about 400 events each year. Stemgas Publishing Co. has it — plus catalog of books, manuals and how-to-do-it data on your hobby. Write Stemgas, Box 328, Lancaster, Pa. 17603 for catalog.

WISCONSIN FIRSTS

Wisconsin is No. 1 nationally in production of milk and cheese. It was the first state to pass a workmen's compensation law, first state with statewide primary election law, and first with an income tax law that worked.

FROGS THAT WARBLE

When frogs warble, they herald rain.

Zuni Indian saying

LEARN ABOUT BIRDS

If you like birds, you'll want to know about the correspondence course offered through Cornell University. Its "Seminars in Ornithology" are written by specialists and well illustrated. For more information write to Laboratory of Ornithology, Cornell University, 159 Sapsucker Woods Road, Ithaca, N.Y. 14850.

LEAKY GARDEN HOSES

You can lose a lot of water — and money — through a leaky garden hose. To prevent leaks, start when you buy by choosing a hose made of high-grade rubber and laminated filament. Diameter of 5/8 inch is good for a medium-sized lawn. Give the hose care — don't put sharp bends in it for long periods; don't keep running your car or tractor over it unprotected; drain it and coil it and store it away in a safe place for winter.

HORSES AS FORECASTERS

If horses stretch out their necks and sniff the air, rain will ensue.

If horses start more than ordinary, and are restless and uneasy, or if they assemble in the corner of a field with heads to leeward (toward shelter), expect rain.

When horses and mules are very lively without apparent cause, look for chill.

June Lore

TREES FOR SHADE

If you wish to plant trees to provide summer shade for your property, give preference to maples and other trees with full crowns.

What you want is trees with high branches that permit visibility on normal eye level, and do not block the approach of cooling breezes.

Evergreens provide best protection in windbreak planting — for times when temperatures are down and wintry winds are blowing. Evergreens, with their cone-shaped crowns, throw little summer shade on walls and roofs. Their low branches lessen visibility and hold off the breezes. If you plant them in the wrong place, they may hold out winter sun's warmth.

Maximum tree shade is provided through group planting, but the roof need not be totally shaded. One study shows that with 20% or more shade during the entire day, cost of air conditioning can be cut appreciably.

MAIL GARDEN COURSES

Pennsylvania State College offers 108 subjects in correspondence courses for those interested in gardening and related subjects.

Mail instruction has been conducted annually since 1892, enrolling over 265,000 persons from all 50 states, and over 30 nations. Course fees run mainly under $10, plus handling costs.

For information on courses, write to Pennsylvania State University, 307 Agricultural Administration Bldg., University Park, Pa. 16802.

BEAN EXPRESSIONS

"Full of beans" — vigorous, energetic.

"Spill the beans" — give away a secret, prematurely.

"Bean Town" — old nickname for Boston, known for its baked beans.

"Beanie" — a small cap, sometimes the type that freshmen at college formerly wore.

"He doesn't know beans about anything" — he's ignorant.

"Beanball" — a throw by a baseball pitcher which intentionally hits a batter on the head.

"We're not playing beanbag" — this is serious business, as in the expression, "We're playing hardball."

"Beanery" — an inferior place to eat.

"Beano" — sometimes used instead of "Bingo," a popular game of chance.

"Beanpole" — a tall skinny person.

"Bean" — the head; "He's off his bean" — he's mentally ill.

"Old bean" — an expression of familiarity.

OBSERVE THE HYACINTH

The name of the hyacinth comes from Greek mythology, along with a legend about a young man that will make you take a second look the next time you examine this colorful flower.

Legend says that Hyacinthus and Apollo were throwing the discus, and as Hyacinthus ran forward to reach for one of Apollo's throws, the discus hit him on the head, wounding him fatally. From his blood sprang a flower, which we today call hyacinth. Bullfinch relates in his Mythology that Apollo, to confer greater honor, "marked the petals with his sorrow, and inscribed, 'Ah! Ah!' upon them, as we see to this day."

ANCIENT NAME FOR A TEMPEST

Euroclydon, a tempestuous wind in the Mediterranean, was mentioned by name in the Bible (Acts XXVII 14). It is said to veer east northeast.

FOGS AS FORECASTERS

Fogs are a sign of change.

Fog from seaward, fair weather; fog from landward, rain (New England).

In the Mississippi Valley, when fogs occur in August, expect fever and ague the following fall.

(Our Almanac of 1884 appeared when one-room schools were the rule, but these ideas apply to today's bigger schools as well.)

HOW TO HAVE A BAD SCHOOL

1st. Elect the most ignorant, bigoted, close-fisted old fogies for directors.

2nd. Employ the cheapest teacher you can get, regardless of qualifications, reputation or experience.

3rd. Find all the fault you can with the teacher, and tell everybody; especially let the pupils hear it.

4th. When you hear a bad report about the teacher or the school, circulate it as fast as you can.

5th. Never visit the school to encourage the teacher.

6th. Take close notice to what seems to go wrong (quiz the scholars) and tell everybody about it except the teacher.

7th. Never advise the children to obey the teacher, and when one is punished, rush to the school room before your passion is cooled, and give the teacher a hearing in the matter in the presence of the school, or allow your children to insult the teacher without reproving them.

8th. Be indifferent about sending your children too regularly, and if they make slow progress, blame the teacher.

9th. Never have your child study at home, or inquire about their progress.

10th. Do not be concerned whether they have the necessary books.

In general, conduct your school on the cheapest possible plan, and let your chief concern be to find fault and devise ways of retrenchment.

If these rules are faithfully carried out, you are not likely to fail in having a bad school; to see your children grow up vicious and ignorant or look back with bitter censure on their parents for robbing them of their birth-right; you will see the morals of their district degenerate, and decent and enterprising people move out of it.

BIRDS AS FORECASTERS

When the thrush sings at sunset, a fair day will follow.

If the cuckoo does not cease singing at midsummer, corn will be dear.

Magpies flying three or four together and uttering brash cries predict windy weather.

RAIN HAS HIGH VALUE

"It seems to me those TV weather people who speak of rain as a dire calamity are just not gardeners," says Patricia O. LaLand, cultural editor in the press bureau of Colonial Williamsburg.

She's got a good point. Rather than merely welcoming promise of sunshine as the only "good" weather news, we should realistically appreciate the value of rain to our gardens, farms, rivers, lakes, ponds and water supplies in general.

Without rain, our crops would curl up and die. Our soil would be baked as hard as asphalt. None of us could survive.

Every inch of rain is a blessing from heaven, when it falls at the right time, at the right place, in appropriate quantities. Too much is no good, of course. But no thinking gardener or farmer keeps telling rain to go away and come again some other day.

We all delight in sunny skies and mild temperatures for our picnics, outdoor weddings, strawberry festivals, open air concerts, and tractor exhibitions. But we can't always have it both ways — clear for our parties, and wet for our flowers, vegetables and trees.

Rain is every bit as glorious as sun. Some folks are so happy with the arrival of long-overdue showers that they go walking in them. They jump with joy — for very often, a steady, soaking rain can be worth a million dollars.

August Lore

TRY A LIVING BOUQUET

Flowers were important to the people of Victorian days, and many gardeners today are following their lead by planting flower boxes or baskets with living plants.

It's part of the container gardening revolution — mixing many plants in one pot or box for a full summer of floral abundance.

Flowers of different shapes, sizes and forms are used. Choose some plants that grow tall, surrounded by some that are shorter, and fill out with some varieties to tumble over the side of the container, which can be a hanging basket as well as one which sits on a table or a patio.

Try mixing leaf textures and colors. Ideas: The silver-white foliage of dusty miller; the feathery thread of lotus vine floating downward from the mass above; the widely varied foliage colors and patterns of the coleus.

Here are more suggestions: Pansies in blue, purple and white; browalia, blue and white; felicia amelliodes (Blue Marguerite), a blue flower with a yellow eye; ageratum, blue; vinca, pink; lobelia, blue, white or soft red; verbena, zinnias, begonias, marigolds, snapdragon, torenia, petunia, impatiens, or cinerarias; for height, snapdragons, geraniums, Shasta daisies, or taller African marigolds.

The container should be decorative as well . . . just so it will hold the necessary amount of soil. Plenty of water is required — daily in hot, windy weather. Fertilizer can be fed in on a slow release basis, following manufacturer's instructions, or in soluble form as you water.

You can make your own color combinations — a Victorian living bouquet can be an expression of your personality If you don't like the way the colors look together, do better on your next venture.

If the tub or box is to be viewed from all sides, put the tallest plants in the center; if it will be seen from the front only, put the big plants in the rear. If you use a hanging basket, be mindful that it will be viewed from below. Overplant rather than underplant; it's easier to remove than insert, once the bouquet is blooming.

DON'T BRING UP HEMP

In the house of the hanged man, one does not talk about rope.

Spanish proverb

Living Victorian bouquet

AUGUST'S NAME

August was formerly the sixth month in the Roman calendar, and was thus known as Sextilis. Augustus, the first Roman emperor, renamed it in his own honor in 8 B.C., because it was his "lucky month."

MARATHON'S LENGTH

The standard length of a marathon race is 26 miles and 385 yards. The race became part of the modern Olympic games in 1896. The standard of length was set in 1924. The most famous marathon in the U.S. is the Boston, run on each Patriot's Day — the third Monday in April.

SHEEP AS FORECASTERS

If old sheep turn their backs toward the wind, and remain so for some time, wet and windy weather is coming.

If sheep gambol or fight, or retire to shelter, it presages a change in the weather.

Old sheep are said to eat greedily before a storm, and sparingly before a thaw.

FLOWER SEEDS CAN SAVE $$$

While money does not grow on trees, savings can blossom in your garden if you grow your own cut flowers.

The National Garden Bureau reports a limited census comparing purchased flowers to flowers grown in home gardens from seed.

One snapdragon stem was reported to cost $1.18, while a packet of seeds would average $1.14 — and the packet keeps producing flowers. Other findings: Carnations averaged 92 cents a stem, vs. 75 cents for an average packet of seeds, and daisies, $1.71 a bunch, 71 cents for a pack of seeds.

END OF THE SEASON

Good gardeners know that end of the growing season is just the start of further chores which can add to the quality of the seasons to come:

Clean up the garden area so that it is neat and tidy . . . Add to your compost pile . . . Plant cover crops and provide winter protection for the soil . . . Clean and repair your tools . . . Make sure produce is stored for future use . . . Review and update soil test reports . . . Chart successes and failures of the season past . . . Compare results with other gardeners . . . Write down your plans for next year.

HOCUS-POCUS ON ARTHRITIS

Beware of quacks who promise to help you cure arthritis. A government brochure written by Annabel Hech tells you:

"Over 31 million Americans suffer from arthritis and related diseases. Unfortunately medical science has not yet found a cure for this oftentimes crippling ailment. Many arthritics, for whom pain is a fact of daily existence, have turned to quack drugs and devices that promise to do the impossible, sometimes with tragic consequences."

Writer Hech warns:

"You can sit in an abandoned radium mine, hang a Vryllium tube on your lapel, bury yourself up to the neck in horse manure, swill Dr. Fenby's Formula X, or take a dose of 'Chuei-Fong-Tou-Gen-Wan,' but you will not be able to cure your arthritis.

"It is a distressing fact of life that although there are drugs to relieve the pain and reduce the inflammation of arthritis — plain aspirin being one of the most effective — there is no cure for this painful and oftentimes crippling disease. Yet many arthritis sufferers turn to quack devices and 'miracle' cures, such as those listed above. They do no good and in some cases do considerable harm."

As of 1980, an estimated $950 million was being spent annually by sufferers on useless nostrums, unapproved devices, unnecessary food supplements, and diet boks.

The therapy needed by arthritis sufferers depends on the type being treated, its severity, and the patient's response.

Some very pertinent information, including stories of quacks who have profited from sufferers' misery, can be found in a free booklet, "Hocus-Pocus Applied to Arthritis," a reprint from the FDA Consumer. It is HHS Publication No. (DA) 81-1081; request it from Consumer Information Center, Dept. G. Pueblo, Colo. 81009.

KATYDIDS AS FORECASTERS

When you hear the first katydid, plant corn (in Missouri).

The louder the katydids sing in August, the bigger the blizzards in December.

When the katydids sing over the wheat stubble, frost is six weeks away.

DIONNE QUINTUPLETS

The date of May 28, 1934, has a special significance in Callendar, Ont., Canada. That is the day when Mr. and Mrs. Oliva Dionne became parents to quintuplets — all girls. The five were the first to survive beyond infancy, a new record for quintuplets. The whole world welcomed news from the modest Dionne home.

VACATION

A vacation is a period of leisure during which we take along twice the clothing we need, but only half the money.

FIRST DAILY NEWSPAPER

The first successful newspaper printed in the United States was the Pennsylvania Packet and Daily Advertiser, starting on Sept. 21, 1784, at Philadelphia.

Publishers were John Dunlap and David C. Claypool. It had been founded by Dunlap in 1771 as a weekly. While the British held Philadelphia, during the Revolutionary War, it was printed in Lancaster (September, 1777-July 1778). It was published several times a week before it became a daily.

BAER'S GARDEN NEWSLETTER

For garden news, ideas and folklore, read Baer's Garden Newsletter, published quarterly to add to Almanac information. Send $3.00 for a year's issues to Baer's, Box 328, Lancaster, Pa. 17603.

PLANT BULBS IN FALL

Fall is the only time for planting bulbs for spring flowering.

It should be done before the earth is hardened by frost. Some growers "stick" the soil in advance so that the openings for the bulbs exist — and plant after first frost.

Experienced gardeners also plant most trees, shrubs and perennials in fall in most parts of the country. Valuable tips for these include: Plant as early as material is available; water well after planting; stake trees over one inch caliper; wrap tree trunks to prevent sun scald; mulch after the first hard frost but before the ground freezes.

For a list of where you can buy this fall, write away for "Gardening by Mail: Where to Buy It." Over 400 plants are listed as available. Send a self-addressed business size envelope (No. 10) to Mailorder Association of Nurserymen, Dept. ABA, 210 Cartwright Blvd., Massapequa Park, N.Y. 11762.

PIED PIPER ANNIVERSARY

Legend has it that 700 years ago — in 1284 — a man known as the Pied Piper agreed to rid the town of Hamelin, Germany, of rats and mice by luring them away with the music of his flute. He did the job but the town fathers refused to pay his fee — and he played his flute some more, luring all the children out of town and away forever. The sad story lives on, though the fate of the children was never learned.

1984 AT LAST

George Orwell, in his 1946 book "The Animal Farm," told us that "All animals are equal, but some animals are more equal than others." Three years later he produced his most famous volume, "Nineteen Eighty-Four." It is appropriate in 1984 to read or re-read that book, and ponder on the message given by Orwell. That was his pseudonym; he was Eric Arthur Blair, who died at 47 in 1950.

TREES AS FORECASTERS

When the locust trees blossom, you can safely take off your long underwear.

When beechnuts are plentiful, expect a mild winter.

If the blooms of the dogwood tree are full, expect a cold winter; when the blooms are light, expect a mild winter.

November Lore

PLANT ROSES IN FALL

Planting roses in the fall is very easy — that's the word from All-America Rose Selections, noted as a highly-experienced group of hybridizers, growers and introducers of fine roses. Rules shown here apply to all but most northern areas of U.S.

After you have chosen your rose location, prepare the ground. Get your roses, prune all canes back to 18 inches, trim any damaged roots and plant as soon as possible. All-America award winners bear the emblem on a green and white identification tag attached to each plant.

Rose plants make little or no growth after fall planting, but the soil settles firmly about the roots during the winter and the plants are ready to start active growth with coming of warm weather, long before most spring roses are purchased and planted.

Prepare and fertilize bed. Dig 15" to 18" wide and as deep. Add quart of peat moss mixed with ½ cup of fertilizer. Mix well with soil. Form a mound in center and position rose on mound.

See that roots are spread out naturally and that bud union (swelling at base of stems) is level with ground. Scatter several inches loose soil over roots and firm well with foot.

Fill remainder of hole with water and allow to drain. See that the bud union remains at ground level.

Fill remainder of hole with soil and tamp. The plant is now ready for the mound of soil which will protect it against rapid freezing and thawing during the winter.

All-America Rose Selections

Mound soil around and over plants to a height of 12". Remove protective mound in the spring after growth starts.

TREASURE TIME

"Dost thou love life? Then do not squander time, for that's the stuff life is made of."

Benjamin Franklin

HUMAN HAIR VS. RABBITS

Some people may think it's folklore, but extension agents say with a straight face that your barber or beautician can help you fight rabbits in winter by supplying you with cuttings of their clients' hair.

Fill small mesh bags with the hair and place the bags near plants that you don't want the rabbits to munch upon.

Dried blood is also reported effective — animal blood will suffice. No one is suggesting you make a human sacrifice to scare away Peter Cottontail.

If these do not appeal to you, try hardware cloth — quarter-inch wire mesh — at the bases of trees and shrubs. This will fend off mice and moles; rabbits may not be deterred, so chicken wire with 1-inch mesh is better.

Repellents sprayed from aerosol cans can help, but spraying must be repeated as long as the cold lasts.

Keep your yard clean as another measure, since stacks of brush or leaves can provide shelter for the rascals in between forays against your garden.

If you want to enlist natural aides, welcome owls, hawks and snakes. You can even buy a "fake snake" made of wood, which you can put in a very visible location.

Winter makes your bulbs very attractive to rabbits, squirrels and field mice. Rabbits often girdle valuable young trees when other food supplies are unavailable. Maintain a winter alert or face some nasty spring surprises.

NEW WAY TO PUT IT

The father of three children in college, all at the same time, said he couldn't afford to travel to the 1984 Olympics — he was suffering from an ailment he called maltuition.

CATTLE AS FORECASTERS

When cattle and horses lie with their heads upon the ground, it is a sign of rain.

If the bull leads the herd in going to pasture, rain must be expected; but if he is careless, and allows the cows to precede him, the weather will be uncertain.

When a cow tries to scratch its ear, it means a shower is very near/when it thumps its ribs with its tail, look out for thunder, lightning, hail (New Jersey).

BOOKS YOU CAN HEAR

Persons who are blind, or who suffer physical handicaps, or who cannot read normal printed matter, can learn the contents of books and other publications through braille, or casettes or discs.

One program is operated by the National Library Service for the Blind and Physically Handicapped, a unit of the Library of Congress. It publishes books and magazines in braille, or recorded in discs and casettes

It furnishes best seller books, mysteries, religious literature and other kinds of works. It takes subscriptions for over 70 magazines in braille and recorded format. Special equipment to play the recordings is available on indefinite loan.

If you are interested in works on music, you can obtain them in braille or in recording, plus instructional materials.

Apply to National Library Service for the Blind and Physically Handicapped, the Library of Congress, Washington, D.C. 20542, to ask for an application and name of a library near you which can help you.

If you cannot read normal printed matter and wish information, you can obtain casettes of certain booklets listed in the Consumer Catalog of the Consumer Information Center. Write to Student Services CI, Recording for the Blind, Inc., 215 E. 58th St., New York, N.Y. 10022, or telephone (212) 751-0860.

Subjects include automobiles, children, employment and education, food, gardening, health, housing, money management, travel, and hobbies.

Guide to Sun and Moon Tables

HOROSCOPE HINTS

Aries, Ram, March 22 through April 20 — Known for leadership and foresight; romantic; have tendency to get too many projects going at one time; warning — loaf every so often.

Taurus, Bull, April 21 through May 21 — Good at planning and pushing; make good marital mates; talented songsters; warning — watch those horns in the china shop.

Gemini, Twins, May 22 through June 21 — Have double sets of assets; creative and able to balance work and home life; warning — everything can be double, including trouble.

Cancer, Crab, June 22 through July 23 — Marked by perseverance; enjoy mysteries; tender in love; warning — be careful of caprice.

Leo, Lion, July 24 through August 23 — Roar to the top; brave and strong; warm lovers but not yielding; warning — your mind may travel faster than your legs can take you.

Virgo, Virgin, Aug. 24 through Sept. 23 — Clear thinkers and careful developers; good at critical analysis — can tell a good pie even if they can't bake one; seek meaningful marriage but sometimes miss it; warning — overwork may make naught of thought.

Libra, Scales, Sept. 24 through Oct. 23 — Well balanced; seek beauty and harmony, don't always get it; men are ardent wooers; women make good companions; warning — don't upset the balance!

Scorpio, Scorpion, Oct. 24 through Nov. 22 — Respected for secret force which can be directed for good or bad; keep trust, deal fairly; very attractive to opposite sex; warning — don't flare up out of control.

Sagittarius, Archer, Nov. 23 through Dec. 22 — Adventurous and loyal; enjoy helping others; a choice target for Cupid's arrows but not always lucky in love; warning — avoid foolhardy hazards.

Capricorn, Goat, Dec. 23 through Jan. 20 — Strong of mind and determined; like politics; more inclined to rational than romantic views; warning — keep others informed of your aims.

Aquarius, Water Boy, Jan. 21 through Feb. 19 — Pour forth their energies, eager to aid everyman; independent, unconventional; affectionate marriage partners; warning — relax.

Pisces, Fish, Feb. 20 through March 21 — Generous to all, often able to see the invisible; always in action; devoted to nuptial mates, sometimes suffering for it; warning — don't let troubles bog you down.

Rising and setting of the sun and the moon are shown on the tables on the following six pages.

Tables are for latitudes of these cities:

Northern States

New York City; Pittsburgh, Pa.; Lincoln, Neb.; Salt Lake City, Utah.

Middle States

Washington, D.C.; Cincinnati, O.; Kansas City, Mo.; Carson City, Nev.

Southern States

Charleston, S.C.; Dallas, Tex.; San Diego, Calif.

* * *

To correct further for longitude, add 4 minutes for each degree you are west of standard time meridian, or subtract 4 minutes for each degree you are east of the standard time meridian.

WHAT'S THE DIFFERENCE?

For those who wonder whether jam and jelly and preserves are all the same, here are distinctions from the U.S. Department of Agriculture:

They are similar in that each is preserved using sugar; all are jellied or partly jellied. Differences include the fruit used, ratio of ingredients, and methods of preparation.

Jelly is made using fruit juice. It is clear and firm enough to hold its shape when removed from the jar.

Jam is made from crushed or macerated fruit. Less firm than jelly, it spreads more easily.

Conserves are jams made form a mixture of fruits including citrus. Sometimes nuts or raisins are added.

Preserves are whole fruits or large pieces of fruit in a syrup that varies in thickness.

Marmalades are usually made from pulpy fruits, with skin and pulp suspended in a clear, jellied liquid. For citrus marmalades, the peel is sliced very thin.

Butters are made by cooking fruit pulp with sugar to a consistency which spreads easily.

Jellied fruit products need a balanced ratio of fruit, acid, pectin and sugar for best results.

January 1984

	Eastern	Central	Mountain	Pacific
N.M.	3—12:16M	2—11:16E	2—10:16E	2— 9:16E
F.Q.	11— 4:48M	11— 3:48M	11— 2:48M	11— 1:48M
F.M.	18— 9:05M	18— 8:05M	18— 7:05M	18— 6:05M
L.Q.	24—11:48E	24—10:48E	24— 9:48E	24— 8:48E

	Northern States			Middle States			Southern States		
	Sun Rises H.M.	Sun Sets H.M.	Moon Rises H.M.	Sun Rises H.M.	Sun Sets H.M.	Moon Rises H.M.	Sun Rises H.M.	Sun Sets H.M.	Moon Rises H.M.
1.	7:38	4:29	6:19	7:22	4:45	6:01	7:08	4:59	5:45
2.	7:39	4:29	7:20	7:22	4:46	7:00	7:08	5:00	6:43
3.	7:39	4:30	sets eve.	7:22	4:47	sets eve.	7:08	5:00	sets eve.
4.	7:39	4:31	5:49	7:22	4:48	6:06	7:09	5:01	6:22
5.	7:38	4:32	6:52	7:22	4:48	7:06	7:09	5:02	7:20
6.	7:38	4:33	7:55	7:22	4:49	8:06	7:09	5:03	8:17
7.	7:38	4:34	8:58	7:22	4:50	9:06	7:09	5:04	9:14
8.	7:38	4:36	10:00	7:22	4:51	10:05	7:09	5:05	10:09
9.	7:38	4:37	11:02	7:22	4:52	11:03	7:09	5:06	11:05
10.	7:37	4:38	morn.	7:22	4:53	morn.	7:09	5:06	morn.
11.	7:37	4:39	12:05	7:22	4:54	12:03	7:09	5:07	12:01
12.	7:37	4:40	1:10	7:21	4:55	1:04	7:08	5:08	12:59
13.	7:36	4:41	2:16	7:21	4:56	2:07	7:08	5:09	1:59
14.	7:36	4:43	3:26	7:21	4:57	3:13	7:08	5:10	3:03
15.	7:35	4:44	4:39	7:20	4:59	4:23	7:08	5:11	4:08
16.	7:34	4:45	5:51	7:20	5:00	5:32	7:08	5:12	5:16
17.	7:34	4:46	6:57	7:20	5:01	6:38	7:07	5:13	6:21
18.	7:34	4:48	rises eve.	7:19	5:02	rises eve.	7:07	5:14	rises eve.
19.	7:33	4:49	6:13	7:19	5:03	6:27	7:07	5:15	6:40
20.	7:32	4:50	7:34	7:18	5:04	7:45	7:06	5:16	7:54
21.	7:31	4:52	8:55	7:18	5:05	9:01	7:06	5:17	9:06
22.	7:31	4:53	10:13	7:17	5:06	10:15	7:05	5:18	10:16
23.	7:30	4:54	11:28	7:16	5:08	11:25	7:05	5:19	11:24
24.	7:29	4:56	morn.	7:16	5:09	morn.	7:05	5:20	morn.
25.	7:28	4:57	12:42	7:15	5:10	12:34	7:04	5:21	12:29
26.	7:27	4:58	1:54	7:14	5:11	1:43	7:03	5:22	1:40
27.	7:26	5:00	3:04	7:14	5:12	2:50	7:03	5:23	2:37
28.	7:25	5:01	4:11	7:13	5:14	3:54	7:02	5:24	3:39
29.	7:24	5:03	5:14	7:12	5:15	4:55	7:02	5:25	4:38
30.	7:23	5:04	6:09	7:11	5:16	5:49	7:01	5:26	5:32
31.	7:22	5:06	6:56	7:10	5:17	6:37	7:00	5:27	6:21

February 1984

	Eastern	Central	Mountain	Pacific
N.M.	1— 6:46E	1— 5:46E	1— 4:46E	1— 3:46E
F.Q.	9—11:00E	9—10:00E	9— 9:00E	9— 8:00E
F.M.	16— 7:41E	16— 6:41E	16— 5:41E	16— 4:41E
L.Q.	23—12:12E	23—11:12M	23—10:12M	23— 9:12M

	Northern States			Middle States			Southern States		
	Sun Rises H.M.	Sun Sets H.M.	Moon Sets H.M.	Sun Rises H.M.	Sun Sets H.M.	Moon Sets H.M.	Sun Rises H.M.	Sun Sets H.M.	Moon Sets H.M.
1.	7:21	5:07	sets eve.	7:09	5:18	sets eve.	7:00	5:28	sets eve.
2.	7:20	5:08	5:45	7:08	5:20	5:58	6:59	5:29	6:56
3.	7:19	5:10	6:48	7:07	5:21	6:58	6:58	5:30	7:06
4.	7:17	5:11	7:50	7:06	5:22	7:57	6:57	5:31	8:02
5.	7:16	5:13	8:53	7:05	5:23	8:55	6:56	5:32	8:57
6.	7:15	5:14	9:55	7:04	5:24	9:54	6:56	5:33	9:53
7.	7:13	5:16	10:57	7:03	5:26	10:53	6:55	5:34	10:49
8.	7:12	5:17	morn.	7:02	5:27	11:54	6:54	5:35	11:48
9.	7:11	5:18	12:02	7:01	5:28	morn.	6:53	5:36	morn.
10.	7:09	5:20	1:09	7:00	5:29	12:58	6:52	5:37	12:48
11.	7:08	5:21	2:19	6:59	5:30	2:03	6:51	5:38	1:51
12.	7:07	5:23	3:28	6:58	5:32	3:10	5:50	5:39	2:56
13.	7:05	5:24	4:36	6:56	5:33	4:16	6:49	5:40	4:00
14.	7:04	5:26	5:38	6:55	5:34	5:18	6:48	5:41	5:01
15.	7:02	5:27	6:29	6:54	5:35	6:12	6:47	5:42	5:57
16.	7:01	5:28	rises eve.	6:53	5:36	rises eve.	6:46	5:43	rises eve.
17.	6:59	5:30	6:26	6:51	5:37	6:34	6:45	5:44	6:41
18.	6:58	5:31	7:49	6:50	5:39	7:52	6:44	5:45	7:55
19.	6:56	5:33	9:08	6:49	5:40	9:07	6:42	5:46	9:06
20.	6:54	5:34	10:26	6:47	5:41	10:20	6:41	5:47	10:15
21.	6:53	5:35	11:41	6:46	5:42	11:31	6:40	5:48	11:23
22.	6:51	5:37	morn.	6:45	5:43	morn.	6:39	5:49	morn.
23.	6:49	5:38	12:54	6:43	5:44	12:40	6:38	5:50	12:29
24.	6:48	5:40	2:04	6:42	5:46	1:47	6:37	5:51	1:33
25.	6:46	5:41	3:09	6:40	5:47	2:50	6:35	5:51	2:34
26.	6:44	5:42	4:07	6:39	5:48	3:47	6:34	5:52	3:30
27.	6:43	5:44	4:56	6:38	5:49	4:36	6:33	5:53	4:20
28.	6:41	5:45	5:36	6:36	5:50	5:19	6:32	5:54	5:04
29.	6:39	5:47	6:10	6:35	5:51	5:55	6:30	5:55	5:42

March 1984

	Eastern	Central	Mountain	Pacific
N.M.	2— 1:31E	2—12:31E	2—11:31M	2—10:31M
F.Q.	10— 1:27E	10—12:27E	10—11:27M	10—10:27M
F.M.	17— 5:10M	17— 4:10M	17— 3:10M	17— 2:10M
L.Q.	24— 2:58M	24— 1:58M	24—12:58M	23—11:58E

	Northern States			Middle States			Southern States		
	Sun Rises H.M.	Sun Sets H.M.	Moon Sets H.M.	Sun Rises H.M.	Sun Sets H.M.	Moon Sets H.M.	Sun Rises H.M.	Sun Sets H.M.	Moon Sets H.M.
1.	6:38	5:48	sets eve.	6:33	5:52	sets eve.	6:29	5:56	sets eve.
2.	6:36	5:49	5:43	6:32	5:53	5:50	6:28	5:57	5:57
3.	6:34	5:51	6:45	6:30	5:54	6:49	6:27	5:58	6:52
4.	6:32	5:52	7:47	6:29	5:55	7:47	6:25	5:59	7:48
5.	6:30	5:53	8:50	6:27	5:57	8:46	6:24	5:59	8:44
6.	6:29	5:55	9:54	6:25	5:58	9:47	6:23	6:00	9:41
7.	6:27	5:56	11:00	6:24	5:59	10:49	6:21	6:01	10:40
8.	6:25	5:57	morn.	6:22	6:00	11:53	6:20	6:02	11:41
9.	6:23	5:59	12:07	6:21	6:01	morn.	6:19	6:03	morn.
10.	6:21	6:00	1:15	6:19	6:02	12:58	6:17	6:04	12:43
11.	6:19	6:01	2:22	6:18	6:03	2:02	6:16	6:05	1:46
12.	6:18	6:02	3:24	6:16	6:04	3:04	6:14	6:05	2:47
13.	6:16	6:04	4:18	6:14	6:05	3:59	6:13	6:06	3:44
14.	6:14	6:05	5:04	6:13	6:06	4:48	6:12	6:07	4:34
15.	6:12	6:06	5:41	6:11	6:07	5:28	6:10	6:08	5:18
16.	6:10	6:08	rises eve.	6:10	6:08	rises eve.	6:09	6:09	rises eve.
17.	6:08	6:09	6:38	6:08	6:09	6:39	6:08	6:10	6:40
18.	6:06	6:10	7:59	6:06	6:10	7:55	6:06	6:10	7:52
19.	6:05	6:12	9:18	6:05	6:11	9:09	6:05	6:11	9:03
20.	6:03	6:13	10:35	6:03	6:12	10:23	6:03	6:12	10:12
21.	6:01	6:14	11:50	6:01	6:13	11:34	6:02	6:13	11:21
22.	5:59	6:15	morn.	6:00	6:14	morn.	6:01	6:14	morn.
23.	5:57	6:17	12:59	5:58	6:15	12:40	5:59	6:14	12:25
24.	5:55	6:18	2:01	5:57	6:16	1:40	5:58	6:15	1:24
25.	5:53	6:19	2:54	5:55	6:17	2:34	5:56	6:16	2:17
26.	5:51	6:21	3:38	5:53	6:18	3:19	5:55	6:17	3:04
27.	5:50	6:22	4:14	5:52	6:20	3:57	5:54	6:18	3:44
28.	5:48	6:23	4:42	5:50	6:21	4:29	5:52	6:18	4:18
29.	5:46	6:24	5:07	5:48	6:22	4:57	5:51	6:19	4:49
30.	5:44	6:26	5:29	5:47	6:23	5:22	5:49	6:20	5:16
31.	5:42	6:27	sets eve.	5:45	6:24	sets eve.	5:48	6:21	sets eve.

April 1984

	Eastern	Central	Mountain	Pacific
N.M.	1— 7:10M	1— 6:10M	1— 5:10M	1— 4:10M
F.Q.	8—11:51E	8—10:51E	8— 9:51E	8— 8:51E
F.M.	15— 2:11E	15— 1:11E	15—12:11E	15—11:11M
L.Q.	22— 7:26E	22— 6:26E	22— 5:26E	22— 4:26E
N.M.	30—10:45E	30— 9:45E	30— 8:45E	30— 7:45E

	Northern States			Middle States			Southern States		
	Sun Rises H.M.	Sun Sets H.M.	Moon Sets H.M.	Sun Rises H.M.	Sun Sets H.M.	Moon Sets H.M.	Sun Rises H.M.	Sun Sets H.M.	Moon Sets H.M.
1.	5:40	6:28		5:44	6:25	6:40	5:46	6:22	6:38
2.	5:38	6:29	7:47	5:42	6:26	7:41	5:45	6:22	7:35
3.	5:36	6:31	8:52	5:40	6:27	8:42	5:44	6:23	8:34
4.	5:35	6:32	9:59	5:39	6:28	9:46	5:42	6:24	9:35
5.	5:33	6:33	11:08	5:37	6:29	10:51	5:41	6:25	10:37
6.	5:31	6:34	morn.	5:36	6:30	11:55	5:40	6:26	11:39
7.	5:29	6:36	12:14	5:34	6:31	morn.	5:38	6:26	morn.
8.	5:27	6:37	1:17	5:32	6:32	12:57	5:37	6:27	12:40
9.	5:25	6:38	2:13	5:31	6:33	1:53	5:36	6:28	1:37
10.	5:24	6:39	3:00	5:29	6:34	2:42	5:34	6:29	2:28
11.	5:22	6:41	3:38	5:28	6:35	3:25	5:33	6:30	3:12
12.	5:20	6:42	4:10	5:26	6:36	4:01	5:32	6:30	3:52
13.	5:18	6:43	4:38	5:25	6:37	4:32	5:30	6:31	4:27
14.	5:17	6:45	5:03	5:23	6:38	5:02	5:29	6:32	5:01
15.	5:15	6:46	rises eve.	5:22	6:39	rises eve.	5:28	6:33	rises eve.
16.	5:13	6:47	8:07	5:20	6:40	7:57	5:26	6:34	7:49
17.	5:11	6:48	9:26	5:19	6:41	9:11	5:25	6:34	8:59
18.	5:10	6:50	10:40	5:17	6:42	10:23	5:24	6:35	10:08
19.	5:08	6:51	11:48	5:16	6:43	11:28	5:23	6:36	11:11
20.	5:06	6:52	morn.	5:14	6:44	morn.	5:21	6:37	morn.
21.	5:05	6:53	12:47	5:13	6:45	12:27	5:20	6:38	12:09
22.	5:03	6:55	1:36	5:12	6:46	1:16	5:19	6:38	12:59
23.	5:01	6:56	2:15	5:10	6:47	1:57	5:18	6:39	1:43
24.	5:00	6:57	2:46	5:09	6:48	2:31	5:17	6:40	2:20
25.	4:58	6:58	3:12	5:07	6:49	3:01	5:15	6:41	2:51
26.	4:56	7:00	3:34	5:06	6:50	3:27	5:14	6:42	3:20
27.	4:55	7:01	3:54	5:05	6:51	3:50	5:13	6:42	3:46
28.	4:53	7:02	4:13	5:03	6:52	4:12	5:12	6:43	4:11
29.	4:52	7:03	sets eve.	5:02	6:53	4:34	5:11	6:44	4:36
30.	4:50	7:05	6:42	5:01	6:54	sets eve.	5:10	6:45	sets eve.

May 1984

	Eastern	Central	Mountain	Pacific
F.Q.	8— 6:50M	8— 5:50M	8— 4:50M	8— 3:50M
F.M.	14—11:29E	14—10:29E	14— 9:29E	14— 8:29E
L.Q.	22—12:45E	22—11:45M	22—10:45M	22— 9:45M
N.M.	30—11:48M	30—10:48M	30— 9:48M	30— 8:48M

	Northern States			Middle States			Southern States		
	Sun Rises H.M.	Sun Sets H.M.	Moon Sets H.M.	Sun Rises H.M.	Sun Sets H.M.	Moon Sets H.M.	Sun Rises H.M.	Sun Sets H.M.	Moon Sets H.M.
1.	4:49	7:06	7:50	5:00	6:55	7:38	5:09	6:46	7:27
2.	4:47	7:07	9:00	4:58	6:56	8:43	5:08	6:46	8:30
3.	4:46	7:08	10:08	4:57	6:57	9:49	5:07	6:47	9:33
4.	4:44	7:10	11:13	4:56	6:58	10:52	5:06	6:48	10:35
5.	4:43	7:11	morn.	4:55	6:59	11:50	5:05	6:49	11:33
6.	4:42	7:12	12:11	4:54	7:00	morn.	5:04	6:50	morn.
7.	4:40	7:13	1:00	4:52	7:01	12:41	5:03	6:51	12:26
8.	4:39	7:14	1:40	4:51	7:02	1:24	5:02	6:51	1:12
9.	4:38	7:16	2:12	4:50	7:03	2:01	5:01	6:52	1:52
10.	4:36	7:17	2:40	4:49	7:04	2:33	5:00	6:53	2:27
11.	4:35	7:18	3:05	4:48	7:05	3:02	4:59	6:54	2:59
12.	4:34	7:19	3:29	4:47	7:06	3:30	4:58	6:55	3:31
13.	4:33	7:20	3:53	4:46	7:07	3:58	4:58	6:55	4:03
14.	4:32	7:22	rises eve.	4:45	7:08	rises eve.	4:57	6:56	rises eve.
15.	4:30	7:23	8:16	4:44	7:09	7:59	4:56	6:57	7:46
16.	4:29	7:24	9:29	4:43	7:10	9:10	4:55	6:58	8:53
17.	4:28	7:25	10:33	4:43	7:11	10:13	4:55	6:58	9:56
18.	4:27	7:26	11:26	4:42	7:11	11:08	4:54	6:59	10:50
19.	4:26	7:27	morn.	4:41	7:12	11:53	4:53	7:00	11:40
20.	4:25	7:28	12:04	4:40	7:13	morn.	4:53	7:01	morn.
21.	4:24	7:29	12:40	4:39	7:14	12:31	4:52	7:01	12:28
22.	4:23	7:30	1:10	4:39	7:15	1:03	4:51	7:02	12:51
23.	4:22	7:31	1:35	4:38	7:16	1:29	4:51	7:03	1:22
24.	4:22	7:32	1:57	4:37	7:17	1:54	4:50	7:04	1:48
25.	4:21	7:33	2:17	4:37	7:18	2:16	4:50	7:04	2:14
26.	4:20	7:34	2:37	4:36	7:18	2:38	4:49	7:05	2:39
27.	4:19	7:35	2:58	4:35	7:19	3:01	4:49	7:06	3:05
28.	4:19	7:36	3:21	4:35	7:20	3:26	4:48	7:06	3:32
29.	4:18	7:37	3:47	4:34	7:21	3:54	4:48	7:07	4:04
30.	4:17	7:38	4:19	4:34	7:22	4:27	4:48	7:08	4:41
31.	4:17	7:39	sets eve.	4:33	7:22	sets eve.	4:47	7:08	sets eve.

June 1984

	Eastern	Central	Mountain	Pacific
F.Q.	6—11:42M	6—10:42M	6— 9:42M	6— 8:42M
F.M.	13— 9:42M	13— 8:42M	13— 7:42M	13— 6:42M
L.Q.	21— 6:10M	21— 5:10M	21— 4:10M	21— 3:10M
N.M.	28—10:18E	28— 9:18E	28— 8:18E	28— 7:18E

	Northern States			Middle States			Southern States		
	Sun Rises H.M.	Sun Sets H.M.	Moon Sets H.M.	Sun Rises H.M.	Sun Sets H.M.	Moon Sets H.M.	Sun Rises H.M.	Sun Sets H.M.	Moon Sets H.M.
1.	4:16	7:40	10:05	4:33	7:23	9:45	4:47	7:09	9:27
2.	4:16	7:41	10:58	4:33	7:24	10:39	4:47	7:09	10:22
3.	4:15	7:41	11:42	4:32	7:24	11:25	4:46	7:10	11:11
4.	4:15	7:42	morn.	4:32	7:25	morn.	4:46	7:11	11:53
5.	4:14	7:43	12:16	4:32	7:26	12:03	4:46	7:11	morn.
6.	4:14	7:44	12:45	4:31	7:26	12:36	4:46	7:12	12:29
7.	4:14	7:44	1:10	4:31	7:27	1:06	4:46	7:12	1:01
8.	4:13	7:45	1:33	4:31	7:27	1:33	4:45	7:13	1:33
9.	4:13	7:45	1:56	4:31	7:28	2:00	4:45	7:13	2:03
10.	4:13	7:46	2:20	4:31	7:28	2:28	4:45	7:14	2:35
11.	4:13	7:47	2:48	4:30	7:29	2:59	4:45	7:14	3:10
12.	4:13	7:47	3:20	4:30	7:29	3:36	4:45	7:15	3:49
13.	4:12	7:48	4:00	4:30	7:30	4:18	4:45	7:15	4:34
14.	4:12	7:48	rises eve.	4:30	7:30	rises eve.	4:45	7:15	rises eve.
15.	4:12	7:49	10:06	4:30	7:31	9:46	4:45	7:16	9:29
16.	4:12	7:49	10:45	4:30	7:31	10:28	4:45	7:16	10:13
17.	4:12	7:49	11:16	4:31	7:31	11:02	4:46	7:16	10:50
18.	4:13	7:50	11:41	4:31	7:32	11:30	4:46	7:17	11:22
19.	4:13	7:50	morn.	4:31	7:32	morn.	4:46	7:17	11:49
20.	4:13	7:50	12:03	4:31	7:32	11:56	4:46	7:17	morn.
21.	4:13	7:50	12:23	4:31	7:32	12:19	4:46	7:17	12:15
22.	4:13	7:51	12:41	4:31	7:33	12:41	4:47	7:17	12:40
23.	4:14	7:51	1:00	4:32	7:33	1:03	4:47	7:18	1:05
24.	4:14	7:51	1:20	4:32	7:33	1:26	4:47	7:18	1:32
25.	4:14	7:51	1:42	4:32	7:33	1:53	4:47	7:18	2:02
26.	4:15	7:51	2:10	4:33	7:33	2:24	4:48	7:18	2:36
27.	4:15	7:51	2:45	4:33	7:33	3:02	4:48	7:18	3:17
28.	4:16	7:51	3:29	4:33	7:33	3:48	4:48	7:18	4:05
29.	4:16	7:51	4:24	4:34	7:33	sets eve.	4:49	7:18	sets eve.
30.	4:17	7:51	sets eve.	4:34	7:33	9:21	4:49	7:18	9:06

July 1984

	Eastern	Central	Mountain	Pacific
F.Q.	5— 4:04E	5— 3:04E	5— 2:04E	5— 1:04E
F.M.	12— 9:20E	12— 8:20E	12— 7:20E	12— 6:20E
L.Q.	20—11:01E	20—10:01E	20— 9:01E	20— 8:01E
N.M.	28— 6:51M	28— 5:51M	28— 4:51M	28— 3:51M

	Northern States			Middle States			Southern States		
	Sun Rises H.M.	Sun Sets H.M.	Moon Sets H.M.	Sun Rises H.M.	Sun Sets H.M.	Moon Sets H.M.	Sun Rises H.M.	Sun Sets H.M.	Moon Sets H.M.
1.	4:17	7:50	10:17	4:35	7:33	10:04	4:50	7:18	9:51
2.	4:18	7:50	10:48	4:35	7:33	10:38	4:50	7:18	10:30
3.	4:18	7:50	11:14	4:36	7:32	11:09	4:51	7:18	11:04
4.	4:19	7:50	11:38	4:36	7:32	11:37	4:51	7:18	11:35
5.	4:20	7:49	morn.	4:37	7:32	morn.	4:52	7:17	morn.
6.	4:20	7:49	12:01	4:38	7:32	12:04	4:52	7:17	12:06
7.	4:21	7:49	12:25	4:38	7:31	12:31	4:53	7:17	12:37
8.	4:22	7:48	12:50	4:39	7:31	1:01	4:53	7:17	1:10
9.	4:23	7:48	1:20	4:40	7:31	1:35	4:54	7:17	1:47
10.	4:23	7:47	1:56	4:40	7:30	2:14	4:54	7:16	2:29
11.	4:24	7:46	2:40	4:41	7:30	3:00	4:55	7:16	3:17
12.	4:25	7:46	3:32	4:42	7:29	3:53	4:56	7:16	4:11
13.	4:26	7:45	rises eve.	4:42	7:29	rises eve.	4:56	7:15	rises eve.
14.	4:27	7:45	9:16	4:43	7:28	9:01	4:57	7:15	8:47
15.	4:28	7:44	9:43	4:44	7:28	9:32	4:57	7:14	9:21
16.	4:29	7:43	10:06	4:45	7:27	9:58	4:58	7:14	9:51
17.	4:30	7:42	10:26	4:45	7:26	10:22	4:59	7:13	10:17
18.	4:31	7:41	10:45	4:46	7:26	10:44	4:59	7:13	10:42
19.	4:32	7:41	11:03	4:47	7:25	11:06	5:00	7:12	11:07
20.	4:33	7:40	11:22	4:48	7:24	11:28	5:01	7:12	11:33
21.	4:34	7:39	11:44	4:49	7:24	11:53	5:01	7:11	morn.
22.	4:35	7:38	morn.	4:50	7:23	morn.	5:02	7:10	12:00
23.	4:36	7:37	12:09	4:50	7:22	12:21	5:03	7:10	12:32
24.	4:37	7:36	12:39	4:51	7:21	12:55	5:04	7:09	1:09
25.	4:38	7:35	1:18	4:52	7:20	1:37	5:04	7:08	1:53
26.	4:39	7:34	2:08	4:53	7:19	2:28	5:05	7:08	2:46
27.	4:40	7:32	3:10	4:54	7:19	3:31	5:06	7:07	3:48
28.	4:41	7:31	4:24	4:55	7:18	sets eve.	5:06	7:06	sets eve.
29.	4:42	7:30	sets eve.	4:56	7:17	8:36	5:07	7:05	8:26
30.	4:43	7:29	9:17	4:57	7:16	9:09	5:08	7:04	9:03
31.	4:44	7:28	9:42	4:58	7:15	9:39	5:09	7:03	9:36

August 1984

	Eastern	Central	Mountain	Pacific
F.Q.	3— 9:33E	3— 8:33E	3— 7:33E	3— 6:33E
F.M.	11—10:43M	11— 9:43M	11— 8:43M	11— 7:43M
L.Q.	19— 2:40E	19— 1:40E	19—12:40E	19—11:40M
N.M.	26— 2:25E	26— 1:25E	26—12:25E	26—11:25M

	Northern States			Middle States			Southern States		
	Sun Rises H.M.	Sun Sets H.M.	Moon Sets H.M.	Sun Rises H.M.	Sun Sets H.M.	Moon Sets H.M.	Sun Rises H.M.	Sun Sets H.M.	Moon Sets H.M.
1.	4:46	7:26	10:05	4:58	7:14	10:07	5:10	7:03	10:07
2.	4:47	7:25	10:29	4:59	7:12	10:34	5:10	7:02	10:39
3.	4:48	7:24	10:54	5:00	7:11	11:03	5:11	7:01	11:11
4.	4:49	7:22	11:22	5:01	7:10	11:36	5:12	7:00	11:47
5.	4:50	7:21	11:56	5:02	7:09	morn.	5:13	6:59	morn.
6.	4:51	7:20	morn.	5:03	7:08	12:13	5:13	6:58	12:27
7.	4:52	7:18	12:37	5:04	7:07	12:56	5:14	6:57	1:13
8.	4:54	7:17	1:26	5:05	7:05	1:47	5:15	6:56	2:04
9.	4:55	7:15	2:23	5:06	7:04	2:44	5:16	6:55	3:00
10.	4:56	7:14	3:25	5:07	7:03	3:44	5:16	6:54	4:00
11.	4:57	7:12	4:30	5:08	7:02	4:46	5:17	6:53	4:59
12.	4:58	7:11	rises eve.	5:09	7:00	rises eve.	5:18	6:52	rises eve.
13.	5:00	7:09	8:31	5:10	6:59	8:25	5:19	6:50	8:19
14.	5:01	7:08	8:50	5:11	6:58	8:47	5:19	6:49	8:45
15.	5:02	7:06	9:08	5:12	6:56	9:09	5:20	6:48	9:09
16.	5:03	7:04	9:26	5:13	6:55	9:31	5:21	6:47	9:34
17.	5:04	7:03	9:46	5:14	6:54	9:53	5:22	6:46	10:00
18.	5:05	7:01	10:08	5:15	6:52	10:20	5:22	6:45	10:30
19.	5:07	7:00	10:36	5:16	6:51	10:50	5:23	6:43	11:03
20.	5:08	6:58	11:10	5:17	6:49	11:28	5:24	6:42	11:43
21.	5:09	6:56	11:54	5:17	6:48	morn.	5:25	6:41	morn.
22.	5:10	6:54	morn.	5:18	6:46	12:14	5:25	6:40	12:30
23.	5:11	6:53	12:50	5:19	6:45	1:11	5:26	6:38	1:27
24.	5:13	6:51	1:57	5:20	6:43	2:17	5:27	6:37	2:33
25.	5:14	6:49	3:14	5:21	6:42	3:31	5:28	6:36	3:45
26.	5:15	6:48	4:36	5:22	6:40	4:48	5:28	6:34	5:00
27.	5:16	6:46	sets eve.	5:23	6:39	sets eve.	5:29	6:33	sets eve.
28.	5:17	6:44	8:06	5:24	6:37	8:06	5:30	6:32	8:06
29.	5:19	6:42	8:30	5:25	6:36	8:34	5:31	6:30	8:38
30.	5:20	6:40	8:55	5:26	6:34	9:03	5:31	6:29	9:11
31.	5:21	6:39	9:23	5:27	6:33	9:36	5:32	6:28	9:46

September 1984

	Eastern	Central	Mountain	Pacific
F.Q.	2— 5:30M	2— 4:30M	2— 3:30M	2— 2:30M
F.M.	10— 2:01M	10— 1:01M	10—12:01M	9—11:01E
L.Q.	18— 4:31M	18— 3:31M	18— 2:31M	18— 1:31M
N.M.	24—10:11E	24— 9:11E	24— 8:11E	24— 7:11E

	Northern States			Middle States			Southern States		
	Sun Rises H.M.	Sun Sets H.M.	Moon Sets H.M.	Sun Rises H.M.	Sun Sets H.M.	Moon Sets H.M.	Sun Rises H.M.	Sun Sets H.M.	Moon Sets H.M.
1.	5:22	6:37	9:56	5:28	6:31	10:12	5:33	6:26	10:26
2.	5:23	6:35	10:35	5:29	6:30	10:54	5:34	6:25	11:10
3.	5:25	6:33	11:22	5:30	6:28	11:42	5:34	6:24	morn.
4.	5:26	6:31	morn.	5:31	6:26	morn.	5:35	6:22	12:00
5.	5:27	6:29	12:16	5:32	6:25	12:37	5:36	6:21	12:55
6.	5:28	6:27	1:17	5:33	6:23	1:37	5:36	6:19	1:52
7.	5:29	6:26	2:22	5:34	6:21	2:38	5:37	6:18	2:52
8.	5:31	6:24	3:27	5:34	6:20	3:41	5:38	6:17	3:51
9.	5:32	6:22	4:32	5:35	6:18	4:42	5:39	6:15	4:50
10.	5:33	6:20	rises eve.	5:36	6:17	rises eve.	5:39	6:14	rises eve.
11.	5:34	6:18	7:14	5:37	6:15	7:13	5:40	6:12	6:48
12.	5:35	6:16	7:31	5:38	6:13	7:35	5:41	6:11	7:13
13.	5:36	6:14	7:50	5:39	6:12	7:57	5:42	6:09	7:37
14.	5:38	6:12	8:12	5:40	6:10	8:22	5:42	6:08	8:03
15.	5:39	6:10	8:37	5:41	6:08	8:50	5:43	6:06	8:30
16.	6:40	6:08	9:07	5:42	6:07	9:24	5:44	6:05	9:02
17.	5:41	6:07	9:46	5:43	6:05	10:05	5:44	6:04	9:39
18.	5:42	6:05	10:35	5:44	6:03	10:56	5:45	6:02	10:21
19.	5:44	6:03	11:35	5:45	6:02	11:56	5:46	6:01	11:13
20.	5:45	6:01	morn.	5:46	6:00	morn.	5:47	5:59	morn.
21.	5:46	5:59	12:47	5:47	5:58	1:05	5:47	5:58	12:13
22.	5:47	5:57	2:06	5:48	5:57	2:20	5:48	5:56	1:21
23.	5:48	5:55	3:28	5:49	5:55	3:36	5:49	5:55	2:32
24.	5:50	5:53	4:51	5:50	5:53	4:56	5:50	5:54	3:46
25.	5:51	5:51	sets eve.	5:51	5:52	sets eve.	5:50	5:52	sets eve.
26.	5:52	5:49	6:54	5:52	5:50	7:00	5:51	5:51	7:06
27.	5:53	5:47	7:21	5:53	5:48	7:32	5:52	5:49	7:42
28.	5:55	5:46	7:52	5:54	5:47	8:07	5:53	5:48	8:20
29.	5:56	5:44	8:30	5:54	5:45	8:49	5:53	5:46	9:04
30.	5:57	5:42	9:15	5:55	5:43	9:36	5:54	5:45	9:53

October 1984

	Eastern	Central	Mountain	Pacific
F.Q.	1— 4:52E	1— 3:52E	1— 2:52E	1— 1:52E
F.M.	9— 6:58E	9— 5:58E	9— 4:58E	9— 3:58E
L.Q.	17— 4:14E	17— 3:14E	17— 2:14E	17— 1:14E
N.M.	24— 7:08M	24— 6:08M	24— 5:08M	24— 4:08M
F.Q.	31— 8:07M	31— 7:07M	31— 6:07M	31— 5:07M

	Northern States			Middle States			Southern States		
	Sun Rises H.M.	Sun Sets H.M.	Moon Sets H.M.	Sun Rises H.M.	Sun Sets H.M.	Moon Sets H.M.	Sun Rises H.M.	Sun Sets H.M.	Moon Sets H.M.
1.	5:58	5:40	10:08	5:56	5:42	10:30	5:55	5:44	10:48
2.	6:00	5:38	11:09	5:57	5:40	11:29	5:56	5:42	11:46
3.	6:01	5:36	morn.	5:58	5:39	morn.	5:56	5:41	morn.
4.	6:02	5:34	12:13	5:59	5:37	12:31	5:57	5:39	12:46
5.	6:03	5:32	1:18	6:00	5:35	1:33	5:58	5:38	1:45
6.	6:04	5:31	2:23	6:01	5:34	2:34	5:59	5:37	2:43
7.	6:06	5:29	3:27	6:02	5:32	3:34	6:00	5:35	3:41
8.	6:07	5:27	4:30	6:03	5:31	4:33	6:00	5:34	4:36
9.	6:08	5:25	5:32	6:04	5:29	5:32	6:01	5:33	5:32
10.	6:10	5:23	rises eve.	6:06	5:28	rises eve.	6:02	5:31	rises eve.
11.	6:11	5:22	6:17	6:07	5:26	6:25	6:03	5:30	6:33
12.	6:12	5:20	6:40	6:08	5:24	6:52	6:04	5:29	7:03
13.	6:13	5:18	7:08	6:09	5:23	7:24	6:04	5:27	7:38
14.	6:15	5:16	7:43	6:10	5:21	8:02	6:05	5:26	8:19
15.	6:16	5:15	8:28	6:11	5:20	8:49	6:06	5:25	9:06
16.	6:17	5:13	9:23	6:12	5:18	9:44	6:07	5:23	10:02
17.	6:19	5:11	10:29	6:13	5:17	10:48	6:08	5:22	11:05
18.	6:20	5:09	11:42	6:14	5:16	11:58	6:09	5:21	morn.
19.	6:21	5:08	morn.	6:15	5:14	morn.	6:10	5:20	12:13
20.	6:23	5:06	1:00	6:16	5:13	1:13	6:10	5:18	1:23
21.	6:24	5:04	2:21	6:17	5:11	2:29	6:11	5:17	2:35
22.	6:25	5:03	3:42	6:18	5:10	3:45	6:12	5:16	3:47
23.	6:27	5:01	5:03	6:19	5:09	5:01	6:13	5:15	5:00
24.	6:28	5:00	sets eve.	6:20	5:07	sets eve.	6:14	5:14	sets eve.
25.	6:29	4:58	5:47	6:22	5:06	6:00	6:15	5:13	6:12
26.	6:31	4:56	6:22	6:23	5:05	6:39	6:16	5:12	6:54
27.	6:32	4:55	7:05	6:24	5:03	7:25	6:17	5:10	7:42
28.	6:33	4:53	7:57	6:25	5:02	8:18	6:18	5:09	8:36
29.	6:35	4:52	8:56	6:26	5:01	9:17	6:18	5:08	9:35
30.	6:36	4:50	10:00	6:27	5:00	10:19	6:19	5:07	10:36
31.	6:37	4:49	11:07	6:28	4:58	11:23	6:20.	5:06	11:36

November 1984

	Eastern	Central	Mountain	Pacific
F.M.	8—12:43E	8—11:43M	8—10:43M	8— 9:43M
L.Q.	16— 1:59M	16—12:59M	15—11:59E	15—10:59E
N.M.	22— 5:57E	22— 4:57E	22— 3:57E	22— 2:57E
F.Q.	30— 3:00M	30— 2:00M	30— 1:00M	30—12:00M

	Northern States			Middle States			Southern States		
	Sun Rises H.M.	Sun Sets H.M.	Moon Sets H.M.	Sun Rises H.M.	Sun Sets H.M.	Moon Sets H.M.	Sun Rises H.M.	Sun Sets H.M.	Moon Sets H.M.
1.	6:39	4:48	morn.	6:29	4:57	morn.	6:21	5:05	morn.
2.	6:40	4:46	12:13	6:31	4:56	12:26	6:22	5:04	12:35
3.	6:42	4:45	1:17	6:32	4:55	1:26	6:23	5:03	1:34
4.	6:43	4:43	2:21	6:33	4:54	2:25	6:24	5:03	2:29
5.	6:44	4:42	3:23	6:34	4:53	3:24	6:25	5:02	3:25
6.	6:46	4:41	4:26	6:35	4:52	4:23	6:26	5:01	4:21
7.	6:47	4:40	5:29	6:36	4:51	5:23	6:27	5:00	5:17
8.	6:48	4:38	rises eve.	6:37	4:50	rises eve.	6:28	4:59	rises eve.
9.	6:50	4:37	5:11	6:39	4:49	5:26	6:29	4:58	5:39
10.	6:51	4:36	5:44	6:40	4:48	6:02	6:30	4:58	6:10
11.	6:53	4:35	6:25	6:41	4:47	6:46	6:31	4:57	6:25
12.	6:54	4:34	7:17	6:42	4:46	7:38	6:32	4:56	7:09
13.	6:55	4:33	8:19	6:43	4:45	8:40	6:33	4:55	8:57
14.	6:57	4:32	9:29	6:44	4:44	9:47	6:34	4:55	10:02
15.	6:58	4:31	10:44	6:45	4:43	10:58	6:35	4:54	11:10
16.	6:59	4:30	morn.	6:47	4:43	morn.	6:36	4:54	morn.
17.	7:01	4:29	12:01	6:48	4:42	12:11	6:37	4:53	12:19
18.	7:02	4:28	1:19	6:49	4:41	1:24	6:38	4:52	1:28
19.	7:03	4:27	2:37	6:50	4:40	2:38	6:39	4:52	2:38
20.	7:05	4:26	3:57	6:51	4:40	3:53	6:40	4:51	3:49
21.	7:06	4:25	5:17	6:52	4:39	5:08	6:41	4:51	5:00
22.	7:07	4:25	6:39	6:53	4:39	6:25	6:42	4:51	6:14
23.	7:09	4:24	sets eve.	6:54	4:38	sets eve.	6:42	4:50	sets eve.
24.	7:10	4:23	5:47	6:56	4:38	6:03	6:43	4:50	6:20
25.	7:11	4:23	7:02	6:57	4:37	7:00	6:44	4:50	7:18
26.	7:12	4:22	7:43	6:58	4:37	8:04	6:45	4:49	8:20
27.	7:14	4:21	8:50	6:59	4:36	9:08	6:46	4:49	9:23
28.	7:15	4:21	9:58	7:00	4:36	10:12	6:47	4:49	10:24
29.	7:16	4:20	11:05	7:01	4:36	11:15	6:48	4:49	11:23
30.	7:17	4:20	morn.	7:02	4:35	morn.	6:49	4:48	morn.

December 1984

	Eastern	Central	Mountain	Pacific
F.M.	8— 5:53M	8— 4:53M	8— 3:53M	8— 2:53M
L.Q.	15—10:25M	15— 9:25M	15— 8:25M	15— 7:25M
N.M.	22— 6:47M	22— 5:47M	22— 4:47M	22— 3:47M
F.Q.	30—12:27M	29—11:27E	29—10:27E	29— 9:27E

	Northern States			Middle States			Southern States		
	Sun Rises H.M.	Sun Sets H.M.	Moon Sets H.M.	Sun Rises H.M.	Sun Sets H.M.	Moon Sets H.M.	Sun Rises H.M.	Sun Sets H.M.	Moon Sets H.M.
1.	7:18	4:20	12:09	7:03	4:35	12:15	6:50	4:48	12:21
2.	7:19	4:19	1:12	7:04	4:35	1:14	6:51	4:48	1:16
3.	7:21	4:19	2:14	7:05	4:35	2:12	6:52	4:48	2:12
4.	7:22	4:19	3:17	7:06	4:35	3:12	6:52	4:48	3:08
5.	7:23	4:19	4:21	7:07	4:35	4:12	6:53	4:48	4:05
6.	7:24	4:18	5:28	7:08	4:34	5:15	6:54	4:48	5:05
7.	7:25	4:18	6:35	7:09	4:34	6:20	6:55	4:48	6:06
8.	7:26	4:18	rises eve.	7:09	4:34	rises eve.	6:56	4:48	rises eve.
9.	7:27	4:18	5:11	7:10	4:35	5:33	6:57	4:48	5:50
10.	7:28	4:18	6:11	7:11	4:35	6:19	6:57	4:48	6:49
11.	7:28	4:18	7:20	7:12	4:35	6:51	6:58	4:49	7:54
12.	7:29	4:18	8:34	7:13	4:35	8:49	6:59	4:49	9:02
13.	7:30	4:18	9:50	7:14	4:35	10:01	6:59	4:49	10:10
14.	7:31	4:19	11:06	7:14	4:35	11:13	7:00	4:49	11:19
15.	7:32	4:19	morn.	7:15	4:36	morn.	7:01	4:50	morn.
16.	7:32	4:19	12:22	7:16	4:36	12:25	7:01	4:50	12:26
17.	7:33	4:20	1:39	7:16	4:36	1:36	7:02	4:50	1:34
18.	7:34	4:20	2:56	7:17	4:37	3:52	7:03	4:51	2:44
19.	7:34	4:20	4:15	7:17	4:37	4:04	7:03	4:51	3:54
20.	7:35	4:21	5:34	7:18	4:38	5:18	7:04	4:52	5:05
21.	7:35	4:21	6:51	7:19	4:38	6:32	7:04	4:52	6:15
22.	7:36	4:22	sets eve.	7:19	4:39	sets eve.	7:05	4:53	sets eve.
23.	7:36	4:22	5:23	7:19	4:39	5:45	7:05	4:53	6:01
24.	7:37	4:23	6:30	7:20	4:40	6:50	7:06	4:54	7:05
25.	7:37	4:24	7:40	7:20	4:40	7:56	7:06	4:55	8:08
26.	7:37	4:24	8:48	7:21	4:41	9:00	7:06	4:55	9:09
27.	7:38	4:25	9:54	7:21	4:42	10:02	7:07	4:56	10:09
28.	7:38	4:26	10:58	7:21	4:42	11:02	7:07	4:56	11:06
29.	7:38	4:27	morn.	7:22	4:43	morn.	7:07	4:57	morn.
30.	7:38	4:27	12:01	7:22	4:44	12:01	7:08	4:58	12:01
31.	7:38	4:28	1:03	7:22	4:45	12:59	7:08	4:59	12:57

BEST BY MAIL

For rates, write National Classified, Box 5, Sarasota, Fla. 33578

MERCHANDISE FOR SALE

FISHERMEN! Try Electric Worm Getters. Safe, easy to use. Complete instructions, $1.50. EWG, 8601 Ferris, Morton Grove, Il. 60053

SENSATIONAL coloured wooden toys. Free catalogue. Norok, 1537(W) Lindley, Phila., Pa. 19141

HORTICULTURE

NEW HYDROPONIC Home Planter. Grows 400 plants annually in 5½ square feet. Build it yourself. Send for fact filled report, $1.00. Satisfaction guaranteed. R.L.Z. Specialty, 400 Davisville Rd., Willow Grove, Pa. 19090

GARDENING Of The Future; Year 2000 And After! Plans and instructions only $3.95. Emerald Solar Grow, 1115 Pine St., Lake in the Hills, Il. 60102

BEER Seed. California Beer Seed. Start package, $10.00. Calvin Perkins, Mathiston, Ms. 39752

SECRET process stimulates plant growth. $18.00/doz. Du-Ma Wholesale, POB 18269(AM), Indianapolis, In. 46218

RARE exotic seeds. Mona's, Rt. 1, Box 165, New Canton, Va. 23123

BOOKS & CATALOGS

FREE Colorado Gift Catalog. Tumbleweeds, Box 461, Morrison, Co. 80465

OPEN House Catalog Featuring Hundreds Of Items. Collectibles, glassware, clocks, jewelry. $1.00 for catalog, refunded with first order. Anderson's Gift House, Rt. 1, Woodriver Sta., Moss, Tn. 38575

FREE! Over 30 quality mail order catalogs, $1.00. Catalog Shopper, 1300(GM) Galaxy Way, Concord, Ca. 94520

OUT Of Print Bookfinder. Send wants. Box 86(AG), Cutten, Ca. 95534

LEGENDARY Gemstones, zodiacs, oils, incense, catalog 37¢. House of Xaro, Box 1933, New Port Richey, Fl. 33552

TREE Trimmers Newsletter, $1. Sample Copy. Box 833(J), Ojai, Ca. 93023

POSTERS framed, matting optional, unbreakable glass. Current Artists. Low, Low Prices. Send Today for full color brochure. Blanche's Posters, 1136 Singleton(vc), Norcross, Ga. 30093

TREASURE of genuine surplus optical parts and instruments. $3. for catalog. T. R. Inc., P.O. Box 65, Mooers, N.Y. 12959

GROW ORIENTAL GOURMET/MEDICINAL MUSHROOMS ON AGRICULTURAL WASTES FOR FUN AND PROFIT

Spawns (seeds) are available in 1-liter (1-qt) unbreakable, autoclavable polypropylene jars. Shiitake (Japanese forest mushroom), Oyster Mushroom, Nameko, Enokitake, Straw Mushroom, as well as Agaricus (common) Mushroom and Chanterelle (a European delight). $9.00-$10.00/jar. For catalog and information, please send $3.00 (deductable from order) to:

OHIO KINOKO COMPANY, P.O. Box 02293
3227-D Riverview Place, Columbus, Ohio 43202

EDUCATION-INSTRUCTION-SCHOOLS

CUT HEATING Costs! Build your own Woodburner — forced-air stove, modern fireplace — from cheap used parts. Handsome, energy-efficient! How To Report, $1.00 (refundable). Famerco, P.O. Box 18471(G), Oklahoma City, Ok. 73154

REPAIR Cuckoo Clocks. Step-by-step. Easy. $3.00. Clockworks, 12910 Orme, Cleveland, Oh. 44125

EARN more money with a High School Diploma. Graduate High School at home! Credits for work/life experience. Low tuition. Budget plans. Toll-free personal tutoring. Accredited diploma. Call 1-800-327-4871 or write: Cambridge Academy, Attn: Roberta, 1553 Arch Ave., Jensen Beach, Fl. 33457

RECIPES

101 TOFU Recipes. Send $3.00 to: Soy World, PO Box 22, Lock Haven, Pa. 17745

CHOCOLATE-Chip Cheesecake, Extraordinary Des-sert, $1.00. Hancock, Box 9013, Washington, D.C. 20003. SASE.

SOAPMAKING! Easy recipes save money! Natural ingredients. $1.00. Country Cottage, Dept. B, Box 223, Roy, Ut. 84067

EGGS Heavenly delight, a new dish and a new sandwich. Recipe, $2.00 and stamped addressed envelope, please. H. Ross, Gypsy Hill, Apt. GO2(A), 500 C St., Staunton, Va. 24401

COFFEE Can Bread! 5 recipes, $2. Irene's, Box 4552, El Paso, Tx. 79914

DISCOVER 10 delicious recipes every month. $5/yr. Subscribe today. Brookside, Box 43036(F), Chicago, 60643

CLASSICS - over 130 Delicious Southern Recipes. Spiral Bound. $6.00, 50¢ postage. Lighthouse, 10810(A) Daisy, Manassas, Va. 22110

How to Turn Your Telephone into a Money Making Machine

Right now your phone is only costing you money—but if you knew how to make it work for you, it could be making you money. Many people have heard about men and women making handsome incomes, via their telephone. But only a few people know exactly how it's done. (only $6.95)

T. H. SMITH
2479 N. 5th St. Milwaukee, Wis.

Washington Monument Capped in 1884

Anyone who believes a national monument can be erected without controversy should take a look at the history of the Washington Monument, which rises majestically above the Mall in the nation's capitol city.

Its height above the floor is 555 feet 5⅛ inches, making it one of the world's tallest monuments constructed of masonry. More than a century passed between the time the idea's suggestion and the setting of the capstone on Dec. 6, 1884, bringing the completion of the work.

Its total cost was $1,187,710; width at base of shaft, 55 feet, 1½ inches; width at top of shaft, 34 feet, 5½ inches; thickness of walls at base of shaft, 15 feet; thickness of walls at top of shaft, 18 inches; depth of foundation, 36 feet, 10 inches; weight of monument, 90,854 tons; sway of monument in wind of 30 miles an hour, 0.125 of an inch.

There are 897 steps on an iron stairway within the monument, with 50 landings. Until 1971 visitors were allowed to walk up these steps. After that they could ride the elevator up, and elect to walk down the steps. Vandalism to memorial stones placed along the stairwell, and some accidents, brought the closing off of the stairwell in 1976.

A steam hoist elevator was used at first; it was replaced by an electric elevator in 1901. An elevator installed in 1959 takes you to the top in 70 seconds.

The story of the monument dates back to 1783, when George Washington was still very much alive and Continental Congress sought a way to honor him as a national hero. Congress delayed, neglecting to take action on any kind of proposal for 50 years — a long time even for Congress. Meanwhile George Washington died in 1799.

Citizens of Washington were organized by leading residents in 1833, as the Washington National Monument Society, to build a monument that was to become a memorial through donations from the public.

Subscriptions totaled $87,000, with interest, by 1847. Robert Mills, noted architect, won the design commission, with a decorated obelisk and a building at its base. Mills' plans were elaborate, but the monument as finally constructed was far different — much simpler.

George Winthrop delivered the address at Masonic ceremonies for the laying of the cornerstone on July 4, 1848. The trowel used for the event was the same used by George Washington at cornerstone ceremonies for the Capitol in 1793.

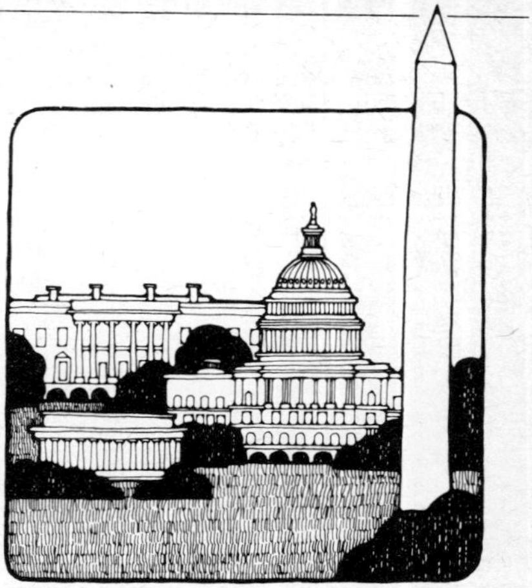

By 1854, masons had reached a height of about 150 feet, when quarreling brought the project to a halt. This acrimony, and then the Civil War, kept the monument an unfinished object into 1876, the year of the Centennial.

President U. S. Grant broke the delay by signing an act giving the government the go-ahead for financing and completion. The Corps of Engineers, in the War Department, was put in charge. Yet that was not enough. Work did not really start again until 1880.

George Winthrop was back to make the dedicatory address — 37 years after the cornerstone laying — on Feb. 21, 1885. And still delay occurred. The public was not allowed to enter until Oct. 9, 1888.

On the interior are 190 memorial stones donated by each of the states, foreign nations and community groups. Windows at the top afford a panoramic view of the city.

Total annual number of visitors is about 1.3 million. The monument is open every day except Dec. 25; winter, 9 a.m.-5 p.m.; summer, 8 a.m.-midnight The monument is administered by the National Park Service, U.S. Department of the Interior.

First Woman Arctic Explorer

Josephine Diebitsch Peary, wife of Robert Edwin Peary, became the first woman member of an Arctic expedition when she accompanied her husband on the "Kite" which sailed from New York on June 6, 1891.

While on the expedition, she gave birth to a daughter, Marie Ahnighnito Peary, farther north than any other Caucasian woman had had a baby. Mrs. Peary later wrote a book, "My Arctic Journal: A Year Among Ice Fields and Eskimos."

Peary became the first person to reach the North Pole, on April 6, 1909.

SAVE ENERGY AT THE DOOR

Some guests start to say good night in the dining room, continue through the parlor and into the vestibule, and then pause at the door to tell you again how much they enjoyed the evening, what wonderful people you are — and, oh, yes, to bring you up-to-date on something they forgot to tell you earlier. The long goodbye can wear down your personal energy, but the holding of the door open, while the winds sneak in and the heat meanders out, can rob you of expensive heating fuel energy. Good rule: Don't tarry at the open door.

WEATHER WARNING

Thou shalt not praise the day 'til night is falling,
However fair its dawn and noon may be;
Oftimes at eventide come storms appalling,
Setting the lightning and the thunder free.

Author unknown

LEARN AT YOUR LIBRARY!

Josephine D. Peary

WHAT HELL IS

Hell is not for flagellation,
Hell is boredom and frustration;
Hell's not fire, as man supposes —
Hell is where you can't grow roses .

Anonymous rosarian

MUNG BEANS

Although much used in the dry seed stage to produce bean sprouts, mung beans are seldom grown in gardens. Most seeds are imported from Asia, but they also can be grown in California and other long-season states.

MARIGOLD COLORS

Marigold flower colors range from very light yellow through orange and reddish brown and now include white. Production of pink, blue or lavender marigolds does not appear likely, yet all sorts of variations are appearing. These include reds midway between scarlet and mahogany, and bicolors that are lavishly decorative. Modern marigold colors are brilliant and do not tend to fade.

RETIREMENT

One wife's definition of retirement: "It's just twice as much husband on half as much money."

A Table for Foretelling the Weather

The following is Dr. Adam Clark's famous table foretelling the kind of weather throughout the Lunations for each year forever. If correct, it will be an invaluable friend to the farmer—and indeed to persons of almost every occupation. Try it:

Table — The table and the accompanying remarks are the result of many years' actual observation, the whole being constructed on a due consideration of the attraction of the sun and moon, in the several positions respecting the earth, and will, by a simple inspection, show the observer what kind of weather will most probably follow the entrance of the moon into any of its quarters, and that so near the truth as to be seldom or never found to fail.

MOON	TIME OF CHANGE	IN SUMMER	IN WINTER
If the New Moon, the First Quarter, the Full Moon, or the Last Quarter, happens	Between midn't and 2 in the morn.	Fair	Hard Frost, unless the wind be S. or W.
	——— 2 and 4, morning,	Cold, with showers	Cold and stormy
	——— 4 and 6, morning,	Rain	Rain
	——— 6 and 8, morning,	Wind and rain	Stormy
	——— 8 and 10, morning,	Changeable	Cold rain, if wind be W. Snow, if E.
	——— 10 and 12, morning,	Frequent showers	Cold and high wind
	At 12, at noon, and 2 P.M.	Very rainy	Snow or rain
	Between 2 and 4, afternoon,	Changeable	Fair and mild
	——— 4 and 6, afternoon,	Fair	Fair
	——— 6 and 8, afternoon,	Fair, if wind N.W.	Fair and Frosty, if wind N. or N.E.
	——— 8 and 10, afternoon,	Rainy, if wind S.W.	Rain or Snow, if S. or S.W.
	——— 10 and midnight,	Rainy, if wind S.W.	Rain or Snow, if S. or S.W.
	——— 12 and 2 A.M.	Fair	Fair and Frosty

Observations — 1. The nearer the time of the Moon's change, Full, and Last Quarter are to *midnight,* the fairer will the weather be during the seven days following. 2. The space for this calculation occupies from 10 at night till 2 next morning. 3. The nearer to *mid-day* or *noon* the phases of the moon happen, the more foul or wet weather may be expected during the next seven days. 4. The space for this calculation occupies from 10 in the forenoon to 2 in the afternoon. These observations refer principally to the Summer, though they affect Spring and Autumn nearly in the same ratio. 5. The Moon's Change, First Quarter, Full and Last Quarter, happening during six of the afternoon hours, *i. e.* from four to ten, may be followed by fair weather; but this is mostly dependent on the wind, as noted in the table. 6. Though the weather, from a variety of irregular causes, is more uncertain in the latter part of autumn, the whole winter, and the beginning of spring, yet in the main, the above observations will apply to those periods also. 7. To prognosticate correctly, especially in those cases where the wind is concerned, the observer should be within sight of a good vane, where the four cardinal points of the heavens are correctly placed. Within this precaution, he will scarcely ever be deceived in depending on the table.

LUCKY MYRTLE

Myrtle has long been considered a shrub or tree of good fortune in England. In some areas, however, there have been conditions. Thus people in Wales would plant myrtle on both sides of the house entrance to keep love and peace within. If it were dug up, however, the love and peace would vanish. Somersetshire legend said that myrtle was by far the luckiest plant for a window box—but would not thrive unless a woman planted it.

STREETS WITH TREE NAMES

William Penn started something in this country when he named streets in Philadelphia for trees. Four of the city's main east-west arteries are Chestnut, Walnut, Spruce and Pine. Pennsylvania, of course, is a combination of his name and "sylvania," a tribute in Latin to its woodlands.

Highlights of the Sky in 1984

Venus is a brilliant object in the morning sky from the beginning of the year until shortly before mid-May, when it becomes too close to the Sun for observation. A few days after mid-July it reappears as an evening star, and can be seen in the evening sky for the rest of the year.

Mars rises soon after midnight at the beginning of the year in Virgo, and moves into Libra in late January. The westward elongation gradually increases until opposition on May 11, when it can be seen throughout the night. The planet enters Scorpius in mid-August, and then moves through Ophiuchus (passing 2°N. of Antares on September 3), Sagittarius, Capricornus and into Aquarius in late December. From late August until the end of the year it can only be seen in the evening sky.

Jupiter rises shortly before sunrise at the beginning of the year and is at opposition on June 29, when it can be seen throughout the night. Its eastward elongation gradually decreases, and from late September it is visible only in the evening sky.

Saturn rises well after midnight at the beginning of the year in Libra and remains in this constellation throughout the year. It is at opposition on May 3, when it can be seen throughout the night. From the end of July until nearly the end of October it can only be seen in the evening sky, and then becomes too close to the Sun for observation. At the end of November it reappears in the morning sky.

MOST IMPORTANT

"Let us never forget that the cultivation of the earth is the most important labor of man."

Daniel Webster

HALLEY'S COMET ON THE WAY

Halley's Comet, which orbits the sun about every 76 years, is "on track" and should be visible to the naked eye beginning in 1986, astronomers report.

First sighting of the comet on its return to our part of the solar system was made in October, 1982, by David C. Jewett and G. Edward Danielson, of the Palomar Observatory near San Diego, operated by the California Institute of Technology.

Scientists of various nations expect to gain more knowledge about the universe on this trip of the comet, named in honor of Dr. Edmund Halley.

Dr. Halley, who was Astronomer Royal of Great Britain, was the first person to note that this comet had appeared in orbit visible from the earth before, a little over 76 years between trips. He made his observations when the comet blazed through the skies in 1682, and stated his belief that is was the same which had been reported upon, in two earlier appearances, in 1531 and 1607.

Dr. Halley made a prediction — that the comet would be seen in 1758 or 1759. He died in 1742, but the comet showed up in 1759 and people started to call it "Halley's Comet." That name has been applied since. The comet last appeared in 1910; it should be at its best visibility in May, 1986.

Halley's Comet is the oldest on which man has made record. Comments on it go back as far as 240 B.C. In ancient days, the brilliant moving light was seen as an omen of ill fortune.

In 1066, it was seen as a good luck sign for William the Conqueror, who defeated King Harold at Hastings. It is represented in a panel of the famous Bayeux Tapestry, which tells the story of the Norman conquest and is on view in a museum in Bayeux, France.

LARGEST LAKE

Lake Superior, one of the Great Lakes, is considered by many experts the largest lake in the world. It covers 31,820 square miles; is 383 miles long and has depths as low as 1,333 feet. Many geographers call the Caspian Sea, in the USSR and Iran, a lake because it is land-locked. It covers 152,239 square miles; is 745 miles long, and has depths to 3,104 feet.

Vegetables to Grow In Containers

Container vegetable gardening is attracting new "customers," because of the way this type of growing can be adapted to modern living conditions.

For the apartment or condominium dweller, containers offer a way in which to raise crops with a minimum of window box, patio or deck space; hanging brackets are also handsome. The city resident with a small yard can utilize it for best effect with containers. The suburbanite may have more growing area, but containers can still be welcome for ease of access.

Make certain you have at least 5 hours of sun a day; sufficient space for the container, and enough air circulation. A bonus is a water supply nearby, with hose or soaker/sprayer nozzle.

Drainage arrangement is necessary; in some cases, a saucer can be enough. Elevate containers where possible to avert root rot from surface water.

Fit container sizes to the plants to be contained. You can make your own or buy wooden, clay, plastic or other kinds. Potting mix should allow drainage, yet retain moisture evenly. Investigate "soilless" commercial potting mixes.

Minister to your plants daily; you can enjoy watching the wonders of growth and development, while scouting to see what will be ready for the table when.

These are just a few tips. Many others can be found in a free booklet offered by the Consumer Information Center, Pueblo, Colo. 81009. Ask for Growing Vegetables In Containers by Kathryn L. Arthurs; No. 615K. Kathryn Arthurs is a garden writer/editor in Palo Alto, Calif.

Here are some of the plants she recommends:

Artichokes — "Green Globe."

Beans — bush forms work best; snap beans "Green Crop," "Tender Crop," "Bush Romano," "Bush Blue Lake," "Royalty (purple pod); lima — "Henderson Bush," "Jackson Wonder Bush."

Brussels sprouts — "Jade Cross," "Long Island Improved."

Cabbage — "Dwarf Morden," "Earliana": Chinese — "Michichli," "Burpee Hybrid."

Carrots — "Danvers Half Long," "Little Finger," "Short & Sweet," "Tiny Sweet."

Chard — any variety in large container.

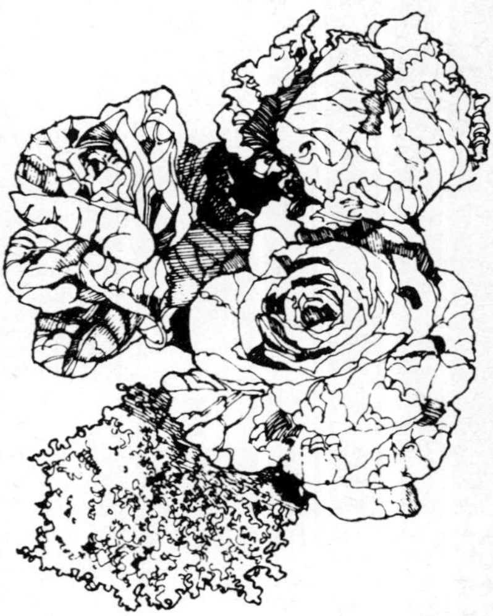

Cucumbers — "Patio Pik," "Pot Luck" developed for containers, good for hanging baskets.

Eggplant — "Modern Midget," "Slim Jim."

Herbs — all do well in containers.

Endive — any variety.

Kale — any variety.

Kohlrabi — any variety; becomes a container conversation piece.

Lettuce — shade can help; any variety.

Melons — midgets are best; "Minnesota Midget," "Yellow Lollipop," "Red Lollipop," "Little Midget."

Okra — Dwarf Green Long Pod," Clemson Spineless," "Red River."

Chives and Scallions — good pot plants

Radishes — all varieties.

Tomatoes — "Tiny Tim," "Small Fry," "Patio Hybrid," "Sugar Lump," "Tumbling Tom" (for hanging baskets), "Stakeless," "Burpee's Pixie," "Salad Top," "Sweet 100," "Toy Boy"

The booklet contains many more ideas. Check your seed supplier for the latest offerings. You'll delight in being able to harvest your crops just before preparation of meals.

Gestation Table

Date of Service	Date Animal Due to Give Birth			
	Mare	*Cow*	*Ewe*	*Sow*
January 1	December 7	October 11	May 31	April 25
January 11	December 17	October 21	June 10	May 5
January 21	December 27	October 31	June 20	May 15
January 31	January 6	November 10	June 30	May 25
February 10	January 16	November 20	July 10	June 4
February 20	January 26	November 30	July 20	June 14
March 2	February 5	December 10	July 30	June 24
March 12	February 15	December 20	August 9	July 4
March 22	February 25	December 30	August 19	July 14
April 1	March 7	January 9	August 29	July 24
April 11	March 17	January 19	September 8	August 3
April 21	March 27	January 29	September 18	August 13
May 1	April 6	February 8	September 28	August 23
May 11	April 16	February 18	October 8	September 2
May 21	April 26	February 28	October 18	September 12
May 31	May 6	March 10	October 28	September 22
June 10	May 16	March 20	November 7	October 2
June 20	May 26	March 30	November 17	October 12
June 30	June 5	April 9	November 27	October 22
July 10	June 15	April 19	December 7	November 1
July 20	June 25	April 29	December 17	November 11
July 30	July 5	May 9	December 27	November 21
August 9	July 15	May 19	January 6	December 1
August 19	July 25	May 29	January 16	December 11
August 29	August 4	June 8	January 26	December 21
September 8	August 14	June 18	February 5	December 31
September 18	August 24	June 28	February 15	January 10
September 28	September 3	July 8	February 25	January 20
October 8	September 13	July 18	March 7	January 30
October 18	September 23	July 28	March 17	February 9
October 28	October 3	August 7	March 27	February 19
November 7	October 13	August 17	April 6	March 1
November 17	October 23	August 27	April 16	March 11
November 27	November 2	September 6	April 26	March 21
December 7	November 12	September 16	May 6	March 31
December 17	November 22	September 26	May 16	April 10
December 27	December 2	October 6	May 26	April 20

GESTATION

Ass	12 Months
Bear	6 Months
Bitch	9 Weeks
Camel	11-12 Months
Cow	9 Months
Cat	8 Weeks
Deer	8 Months
Dromendary	12 Months
Dormouse	31 Days
Elephant	21 Months
Goat	5 Months
Guinea Pig	68 Days
Giraffe	14 Months
Lion	108 Days
Mare	11 Months
Monkey	7 Months
Opossum	26 Days
Rabbit	30 Days
Sheep	5 Months
Sow	16 Weeks
Squirrels and Rats	28 Days
Wolf and Fox	62 Days

Duration and frequency of heat in farm animals in regular condition

	In heat for	If not impregnated, heat will recur in
Mares	5-7 days*	3 to 6 weeks
Cows	2-3 days*	3 to 4 weeks
Ewes	2-3 days	17 to 28 days
Sows	2-4 days	21 days

*Subject to variation

1984 Canadian Weather—1st 4 Months

JANUARY

(Synopsis: Heavier rain and snow West Coast Provinces; seasonal in East, a bit drier Manitoba and Ontario. Temperatures cold Central, near normal West Coast, and changeable in East.)

1- 3 A quick change time. Much warmer Alberta east into Ontario. Rain and snow West Coast. Showery East Provinces.

4- 7 Milder time. Milder everywhere; rain/snow Manitoba to Hudson Bay. Clear Alberta worries farmers. East Coast receives some rain and snow.

8-11 Crazy weather time. West Coast very wet then east into Manitoba, then cold. Cold air from Northwest modified by 9th. More snow East Provinces then cold but clear.

12-15 Constant change time. Polar air in Manitoba and Saskatchewan east into Labrador. Heavy snow East Quebec then into Nova Scotia then colder. Milder and dry Alberta and Saskatchewan; cloudy on Coast.

16-19 Milder days. Pressure centers travel rapidly coast to coast; little temperature change.

20-23 A colder time. Clear Saskatchewan and Manitoba then east. Late heavy rain or snow Nova Scotia north. Rain and snow frequent in West.

24-27 Colder Central Provinces. Heavy snow West Coast into Saskatchewan. Snow east of Hudson Bay. Very cold Manitoba then Western Ontario.

28-31 Heavy precipitation West. Heavy rain or snow British Columbia; mild and dry Alberta and Saskatchewan, moderate in Ontario and Quebec; unsettled Eastern Provinces.

FEBRUARY

(Synopsis: A very cold month, especially in North and Central Provinces, near normal East and West Coast. Rain and snow rather heavy for the month, very heavy in Saskatchewan and Nova Scotia, a bit drier in Manitoba north.)

1- 3 Snow, cold time. Brief but widespread cold and light snow east of Alberta into Newfoundland. Much colder Saskatchewan and Manitoba. Heavy snow West Coast.

4- 7 Milder time. Very cold clear air follows Hudson Bay storm east. Rain/snow both Coasts. Milder by 10th.

8-11 Cold then mild. Very cold, clear air follows storm to Hudson Bay then east. Rain and snow on Coast. Widespread light snow, milder by 10th.

12-15 East snow, West very cold. Continued cold Alberta into Ontario. Snow east of Great Lakes to Atlantic. West Coast much rain and snow.

16-19 Snow showers, cold. Snow showers, cold Alberta-Saskatchewan. Clear, very cold Ontario east, snow by 19th in Eastern Provinces.

20-23 Unsettled days. Milder Saskatchewan and South Manitoba, then colder in Ontario by 23rd. Heavy snow from eastern storm Eastern Quebec, Nova Scotia and Newfoundland.

24-29 Cold.

MARCH

(Synopsis: Cold and wet until near 22nd when milder air brings rain and danger of floods in West. Heavy rain and snow East and Central portions after 22nd. Mild to warm after early cold days Ontario east; continued cold and windy elsewhere.)

1- 3 Warming days. Moderate rain, snow British Columbia and Alberta, light in Saskatchewan.

4- 7 Good snowfall in East. Late prospects of snow from Ontario to Coast. Milder in west. Pacific Coast has heavy rain and snow.

8-11 Cold then warm. Brief cold air interior across Hudson Bay. Rain and snow British Columbia into Manitoba. Squally Eastern Provinces.

12-15 Western storm. Unsettled time. Storm batters West Coast. Cloudiness, light snow showers Ontario east; unsettled Saskatchewan and Manitoba.

16-19 Western storm. Heavy snow inland British Columbia into Manitoba. Severe cold Northwest Territory into Ontario. Lakes storm brings snow Quebec east then colder.

20-23 Major storm in East. After cold days heavy rain and snow storm covers Provinces. Very cold air moves east to Coast. Heavy rain and snow British Columbia into Saskatchewan.

24-27 Hot and cold days. Heavy rain and snow British Columbia brings floods locally. Unsetled Ontario and east then cold. Clouds then rain, milder Saskatchewan into Manitoba.

28-31 Variable but cool. Many areas of rain and snow. Polar cold builds to move south into Hudson Bay region by 30th.

APRIL

(Synopsis: Precipitation heavy in West, moderate elsewhere. Temperature cool West, cold East).

1- 3 Rather cold now. Many light snow showers Ontario east. Clear, frosty, cold Saskatchewan into Manitoba.

4- 7 Warmer and wet. Snow moving storm centers both Coasts. Squalls in Manitoba then clear and cold.

8-11 Heavy rain and snow; milder West. A late storm center Nova Scotia moving north. Broad mild storm center British Columbia into Manitoba. Clear cold Ontario and Quebec.

12-15 Warm West; cold center, stormy East. Everything moves ever so slowly now as deep eastern storm continues north.

16-19 Mild air moves fast. Cloudy and mild British Columbia east into Saskatchewan. Clear, cold air Manitoba moves east to Hudson Bay. Rain and snow Nova Scotia north.

20-23 Soft touch of winter. Rain and snow, colder British Columbia east. Sharp colder Manitoba east to Hudson Bay. Clear cold then rain or snow Quebec east to Coast.

24-27 Very cold, then warm. Vlear, very cold Eastern Provinces then rain or snow Saskatchewan and Manitoba east.

28-30 Warm and cold. Cloudy then rain or snow, even thunderstorms, preceed colder air from Pacific east by 30th.

1984 Canadian Weather — 2nd 4 Months

MAY

(Synopsis: A change to milder days by end of month everywhere; showers and rain excellent for growing crops and gardens but a bit cool Western mountains.)

1- 3 Cold East warmer West. Eastern Provinces colder, warm West. Showers, milder southern Alberta and Saskatchewan; coastal showers British Columbia.

4- 7 Cool wet days. More showers British Columbia east into Manitoba. Clear frosty Ontario into Labrador. Showery on Coast.

8-11 Unsettled days. Rain and snow British Columbia. Unsettled Ontario and Quebec. Rain and snow Nova Scotia. Clear cold Alberta and Saskatchewan, Western Manitoba.

12-15 Wheat land rains. Local showers East and West Coast. Good rain showers Saskatchewan and Manitoba. Unsettled Ontario and Quebec.

16-19 Good rain Center and Eastern Provinces. The Great Plains rain spreads north and east into eastern Manitoba then Ontario and Quebec.

20-23 Cold to warm days. Coast rains continue east into Alberta and Manitoba then clear cooler. Clouds then showers Ontario.

24-27 Great growing weather. Showers and rain then clearing skies, good growing weather.

28-31 Looking good. Clear colder British Columbia east into Alberta. Local showers Eastern Provinces, less in Manitoba and Ontario.

JUNE

(Synopsis: Unusual weather this month: Very cold then warm again, cold and wet. Damage to grains could be serious farther north, less to the south and east.)

1- 3 Rains in West. Beneficial rain and showers British Columbia into Saskatchewan. Unsettled West Manitoba. Clear, cool Eastern Manitoba then Quebec.

4- 7 Cold West, fog East. Large areas from Ontario to Atlantic Coast have rain and fog then clear and cold. Numerous storms and heavier showers in West by 7th.

8-11 Good showers in grain areas. Slow moving shower cells move across grain growing area.

12-15 Plenty of showers. Many showers and thunderstorms west of Manitoba. Many storms break short hot spell by 15th east of Ontario.

16-19 Wet and sassy. Many cells of showers and rain British Columbia east into Saskatchewan. Clear colder Manitoba, also Western Ontario. Unsettled Eastern Ontario and Quebec.

20-23 Unusually cold passage. Clear after rain or snow British Columbia into Alberta. Cold air from Northwest Territory covers Ontario-Quebec moving east. Local showers Saskatchewan-Manitoba.

24-27 Beneficial rain days. Beneficial rains Eastern Alberta into Western Manitoba. Showery Ontario east to the Coast then clear, very cool.

28-30 Clear then showers. Clear days follow showers and storms coast to coast. Temperatures are agreeable.

JULY

(Synopsis: Devastating heat and cold with damaging rains Central Region and near Lakes. Rain local heavy Saskatchewan east into Ontario; fog on East Coast frequent.)

1- 3 Clear cool Saskatchewan and Manitoba. Few clouds British Columbia and Alberta. Rain and showers Maritimes. Clouds Quebec.

4- 7 Little change except clouds then rain Ontario-Quebec moving into Newfoundland and Nova Scotia.

8-11 Good rain showers help crops Saskatchewan-Manitoba into Western Ontario. Part cloudy then showers and rain late in Quebec and the Maritimes. Clear, cool British Columbia.

12-15 Spotty shower cells British Columbia-Manitoba then cooler Pacific air. Part cloudy, warm Ontario-Quebec then clear, warm.

16-19 Unusual cold season in West moves east from Saskatchewan preceeded by very heavy damaging rain storms.

20-23 Temporary relief from past heat in East from Ontario to the Coast as more heat from the South is expected.

24-27 A cool few days Manitoba-Western Ontario then more heat. British Columbia and Alberta remain cool. Rain continues near Great Lakes.

28-31 Many rain showers Quebec and Maritimes. Polar air brings dryness into Saskatchewan West Ontario. Brief showers then clear British Columbia east.

AUGUST

(Synopsis: A very changeable month with periods of cool clear air and then very warm followed by showers and storms that are damaging. Watch the cold air movements.)

1- 3 Heavy rain and wind damage likely from southern storm on Atlantic Coast. Clear, cool Ontario and Quebec.

4- 7 Widespread showers with clear periods everywhere. First of real autumn-like weather.

8-11 Many showers continue from Central Ontario thru the Provinces then clear and cooler. Waves of showers and rain then clear, cool British Columbia continue east.

12-15 Rain and showers continue to delay the harvest East Manitoba into Quebec. Cloudy in East.

16-19 Rain is light Ontario to Coast. Unsettled British Columbia thru Alberta. Rain could be very heavy before clear skies cover Manitoba.

20-23 Slow moving depressions deposit rain and showers Saskatchewan into Quebec. Clear, warm Eastern Provinces then clouds, showers. Clear South, showery North to Central British Columbia and Alberta.

24-27 Very warm air from Great Lakes sweeps east rapidly then clear cool, showers and storms follow. Showers and storms general by 27th.

28-31 Cold air to near freezing in portions of Alberta into Manitoba spreads east later. Clear, mild air, some clouds in mountains of British Columbia, Alberta, also Atlantic Coast.

1984 Canadian Weather—3rd 4 Months

SEPTEMBER

(Synopsis: Early freeze Saskatchewan east into Ontario does serious damage to crops locally. Rain is ample, especially in Eastern Provinces.)

1- 3 Sharp but brief freeze Saskatchewan-Manitoba continues into Ontario then milder. Brief showers British Columbia, then east.

4- 7 Freeze continues in Ontario and Quebec. Clouds then light rain and showers Saskatchewan-Manitoba.

8-11 Cold frosty skies move quickly across Ontario and Quebec then milder winds and showers even to coast. Fog frequent in Maritimes.

12-15 Rain and showers continue in East. Clear, dry, light winds Alberta to Saskatchewan. Unsettled elsewhere.

16-19 Very humid and mild with more storminess Ontario, Quebec moving east then clear, colder. Storm West Coast sends waves of rain east.

20-23 Showers are brief and light as storm center moves from Manitoba toward Nova Scotia. Showers mild on British Columbia Coast.

24-27 Cold polar air spreads east across Ontario. Milder wet air dominant in British Columbia into Saskatchewan.

28-30 Alaskan storm centers bring heavy rain or snow to Pacific Coast east thru Manitoba into Ontario then Quebec.

OCTOBER

(Synopsis: Pacific stormy and very wet, also the Eastern Provinces Nova Scotia north. Temperatures are lower in British Columbia into Manitoba and above to the East.)

1- 3 Rain and snow rather widespread preceding cold air from Northwest Territory. Effective storm on British Columbia Coast and another on Atlantic Coast.

4- 7 Pacific Coast stormy then clear but cold. Variable cloudiness, light rain and showers Manitoba-Western Ontario. Clear cool then milder.

8-11 Pacific air follows brief storm centers east. Rain and showers brief. Clear, cold frosty air farther north.

12-15 Waves of showers then snow but colder follow from West Coast into Manitoba.

16-19 Gales with rain or snow Newfoundland, Nova Scotia. Unsettled Ontario and Quebec. Rain or snow Alberta to Coast.

20-23 Blustery, some rain or snow, cold Ontario-Quebec east to Coast. Clear, cold to Manitoba.

24-27 Less rain and snow Saskatchewan-Manitoba then clear, cold. Rain or snow Ontario east thru Provinces, continued cold.

28-31 Light rain or snow then clear cold British Columbia and Alberta. Cyclonic winds increase in Saskatchewan-Manitoba moving east.

NOVEMBER

(Synopsis: A very cool month with plenty of light snow widespread that is heavy in the mountains.)

1- 3 Cold, frosty clear skies Eastern Maritimes then brief rain and snow soon follows. Rain and snow British Columbia into Saskatchewan.

4- 7 Very cold northwestern Canadian air eased somewhat by ocean storms and fast moving fronts. Snow is widespread in Manitoba.

8-11 Snow British Columbia, Alberta, also Ontario-Quebec. Blizzard likely Labrador then cloudy south thru Nova Scotia.

12-15 Storm from Great Lakes brings extended rain and snow with a brief cold snap then more rain and snow, Ontario east to Coast.

16-19 Great Lakes storm deposits heavy rain and snow then much colder. Clear and cold in British Columbia-Alberta, moves east.

20-23 Cold high pressure from Western Canada keeps Manitoba-Ontario covered by snow. Cloudy Eastern Provinces.

24-27 Heavy storm centers on both Coasts bring many periods of rain or snow. Quite clear and cool Manitoba into Quebec.

28-30 The storms on both Coasts spread snow farther inland. Interior is very cold with zero over Northern Saskatchewan into Quebec.

DECEMBER

(Synopsis: Many rain and snow gales on both Coasts while interior is very cold but drier except in mountains. There is more snow in Ontario than to the West.)

1- 3 Waves of rain and snow from British Columbia dissipate over Ontario. East Coast storm becomes weaker then clear and very cold.

4- 7 Rain and snow frequent into Saskatchewan. Ridge of cold air from Far North keeps Manitoba into Quebec cold as gales subside.

8-11 Again rapid moving pressure centers bring unusual unsettled weather from West to East Coast. Clear periods are short.

12-15 Pacific storms become docile when near Alberta but in cold interior bring many snow squalls. East storm center becomes blizzard.

16-19 Storms from Pacific continue to send snow and rain inland in active centers keeping cold in Northern Provinces.

20-23 Heavy snow in West may set a record as Siberian cold at last covers much of region from Saskatchewan into Eastern Quebec.

24-27 Heavy coast storm. Heavy snow to British Columbia east thru Saskatchewan. Lake storm brings snow to Ontario-Quebec east to Coast.

28-31 Widespread clouds, light snow squalls in all Provinces then cold high pressure clears Manitoba east.

Probable Canadian Weather—1983

SEPTEMBER

1- 3 Autumn close. Showers increase Ontario and Quebec. Cold air poised in central Canada to move south across Manitoba and eastern Saskatchewan. Showers and rain on Atlantic Coast move inland.

4- 7 Showers and storms delay harvest in Saskatchewan, continue east rapidly. Clear warm weather remains near Great Lakes.

8-11 Showers or storms in Saskatchewan and Manitoba dissipate as cold air from North flows south. Rain over Maritimes; clear, frosty skies warm quickly.

12-15 Clear and cold western Provinces. A narrow band of clouds, rain crosses Ontario, Quebec.

16-19 Freeze and thaw. Light rain or snow north Saskatchewan west to Pacific Coast. Increasing danger of gales Nova Scotia north.

20-23 Early Autumn temperatures rise to above normal, showers local, Saskatchewan, Alberta. Unsettled British Columbia.

24-27 Good days. Rain spreads into Manitoba and Ontario. Clear after early fog in Maritimes. Clearing skies British Columbia and Alberta.

28-30 Colder days. Heavier rain and snow Quebec east into Maritimes. Coastal gales decreasing on Pacific coast.

SEPTEMBER SUMMARY: Near normal temperatures and rainfall in Maritimes. Wet and cold Alberta and Saskatchewan. Warmer, a little dry Manitoba into Quebec; normal elsewhere.

OCTOBER

1- 3 Autumn rains, falling leaves. Waves of rain roll across entire Canada. Cold air from British Columbia moves east late in period.

4- 7 West Coast gales into Alberta. Unsettled Saskatchewan and Manitoba. Cold and partly cloudy into Maritimes.

8-11 Milder temperatures West and East, but cold Central. Increasing winds, then gales Nova Scotia, moving north. Rain, snow in British Columbia continue east. Polar cold air.

12-15 Clear, cold air British Columbia into western Ontario. Seasonally cold, some cloudiness in eastern Ontario thru Quebec.

16-19 Some clearing skies after brief showers British Columbia into Manitoba move east. Clear, frost at night Ontario and Quebec.

20-23 Major storm now. Winter comes early. Clear British Columbia. Snow Alberta and western Saskatchewan. Cloudy, then snow Manitoba.

24-27 Stormy, then clear, cold. Almost coast to coast showery. Heavy storm Manitoba and Ontario, then clear, cold with frosts Manitoba east.

28-31 Stormy in West. West Coast storm east into Saskatchewan. Clear, cold in Manitoba and Ontario. Decreasing storminess eastern Quebec and Atlantic Coast.

OCTOBER SUMARY: A cold month from British Columbia into Saskatchewan. Drier and frosty, Manitcba into Ontario. Showery and near normal rain and snow Quebec east.

NOVEMBER

1- 3 It is Autumn. Coast rain and snow British Columbia. Clear, frosty, then clouds Saskatchewan and Manitoba. Scattered showers Quebec.

4- 7 Constant West Coast storm sends waves of clouds, showers east. Clear to partly cloudy eastern Quebec. A dissipating storm northeast of Nova Scotia continues northeast.

8-11 Clear, cold West; clouds East. Clear, cold British Columbia and Alberta. More gales, also rain or snow Maritimes.

12-15 Wet West, cool East. Heavy rain, even snow British Columbia east into Saskatchewan.

16-19 Many showers Manitoba east thru Maritimes, then sharply cloudy. Rain, snow West Coast.

20-23 Cold to warm days. Indian Summer, maybe. Cloudiness, warmer air from south Saskatchewan east thru Ontario. Short session of clear, crisp weather British Columbia to Maritimes.

24-27 Waves of storminess West Coast move east, then colder. Brief storms, then cold Labrador.

28-30 Strong storms West Coast. Odd weather days everywhere because cloudiness is short and sunny time follows. More rain, snow northern British Columbia into Saskatchewan.

NOVEMBER SUMARY: Cold and snow-bearing month in Alberta, Saskatchewan and northern Maritimes. Cold and showery Manitoba into Ontario. Near normal elsewhere.

DECEMBER

1- 3 Only western Quebec to the Maritimes has any snow and snow showers.

4- 7 Warmer days. Rain and snow move inland into Alberta. Less effective in Saskatchewan.

8-11 Decreasing rain West. Cloudy British Columbia, Alberta. Clear Saskatchewan into west Ontario.

12-15 Brrr, it is cold! Sub-zero now into East. Very cold following light snow Manitoba east. Late snow squalls British Columbia.

16-19 Generous snow East, clear cold West. Cloudy coast British Columbia. Clear Alberta east into Manitoba. Snow in Quebec moving into Maritimes. Gales with snow late in Maritimes.

20-23 Cold, then stormy. Snow British Columbia into Saskatchewan spreads east as storm intensifies in Ontario. Cold, clear in Maritimes.

24-27 Snow showers continue in Alberta east across Saskatchewan into Quebec. Fresh southwesterly winds, rather mild in Maritimes.

28-31 Winter snows in British Columbia, Alberta. Cloudy, local snow showers Saskatchewan east into Quebec. Gales continue in eastern Maritimes; clear, cold in western portion.

DECEMBER SUMMARY: Real winter with plenty of snow and few warming days anywhere.

Philosophy from Thoreau

Henry David Thoreau

Henry David Thoreau, who loved the woods, and gardening, and writing, has left a legacy in sage observations, some of which we quote here:

"The outward is only the outside of that which is within. Men are not concealed under habits, but are revealed by them; they are not their true clothes." (In this sense, we think, Thoreau referred to "habits" as clothing.)

"In the long run men hit only what they aim at."

"Read the best books first, or you may not have a chance to read them at all."

"I shall not be as cheap to myself if I see that another values me."

"We are as much as we see."

"The bluebird carries the sky on his back."

"The youth gets together his materials to build a bridge to the moon, or perchance a palace or temple on the earth, and at length the middle-aged man concludes to build a woodshed with them."

"Some circumstantial evidence is very strong, such as when you find a trout in the milk."

"Beware of all enterprises that require new clothes."

"To him whose elastic and vigorous thought keeps pace with the sun, the day is a perpetual morning."

"To be awake is to be alive."

"If a man does not keep pace with his companions, perhaps it is because he hears a different drummer. Let him step to the music he hears, however measured or faraway."

RAISED BED GARDENS

Some gardeners use old railroad ties to contain their raised gardens; others prefer thinner boards placed on edge; all find that little maintenance time is required.

Weeds pull out easily from the porous soil in a raised bed; watering goes rapidly. Virtually no plants are lost to root rot. Rain causes few delays in cultivation and harvesting. If you're new at it, try a small section first; add more in later years.

Michigan Marvels & Minutiae

A Detroit man was awarded $2,067,696 after party-goers threw a couch out the fourteenth floor window of a Detroit hotel in December, 1975, and hit him on the head.

The highest temperature recorded in Michigan was 112° F, at Mio on July 13, 1936; the lowest was -51° F, at Vanderbilt, Feb. 9, 1934.

Michigan national champion tall trees include: American Beech, at Three Oaks, 161 feet high; Black Cherry, 132, Washtenaw County; Thornless Honeylocust, 130, Washtenaw County; Balsam Poplar, 128, Champion, and Red Maple, 125, Armada.

These are but a few of the distinctions and oddities of Michigan, among the thousands to be found in a book called "Michillaneous," compiled by Gary W. Barfknecht, of Davison.

Barfknecht gathered all the information over a period of two years. We read about the book in an Associated Press dispatch in the Lancaster (Pa.) New Era. Barfknecht sent a review copy, and we feel that when word gets around, similar books will appear on all the other 49 states.

Here are some more of his Michigan marvels and minutiae . . .

The Tulip Festival, held each May at Holland, Mich., is one of 27 ethnic festivals in the state; so far as we know, it is the one most publicized nationally.

The Mackinac Bridge, linking Mackinac City and St. Ignace, is the longest suspension type bridge in the world, from cable anchor to cable anchor, and has been designated as one of the seven man-made wonders of the world by the American Institute of Architects.

Monster Vegetables — winners in contests for the largest in the state sponsored by Michigan Natural Resources magazine, were (in 1981) — Beet, 5 lbs., 12 oz.; cabbage, 18 lbs.; carrot, 3 lbs.; celery, 10 lbs., 12 oz.; potato, 4 lbs., 2½ oz.; pumpkin, 224 lbs., 8 oz.; radish, 8 lbs., 14 oz.; tomato, 2 lbs., 12 oz.; zucchini, 20 lbs.

Six poisonous plants (not really limited to Michigan) — Buttercup, all parts of plant are poisonous; acorns, eating too many can gradually poison the kidneys; cherries, twigs and foliage are dangerous; jimson weed, the juice of the prickly, fleshy fruit can kill; nightshade, berries, especially unripe, are poisonous; elderberries, all parts of the plant are toxic.

Michigan's worst winter blizzard, on Jan. 26, 1978, brought up to 2 feet of snow and winds of 60 to 75 miles an hour, with drifts 10 feet high. The storm killed 20 persons and stranded 400,000 along the state's highways.

Symbol of Holland, Mich., Windmill Island Municipal Park. The windmill, "De Zwaan," dates back 200 years to its original site in the Netherlands, near Amsterdam. It was moveed in 1890 to Vinkel, Noord Brabant, and then in recent years to Holland, Mich. The sails are mounted on steel beams 80 feet long, which reach to the height of a 12-story building.

The greatest amount of rain falling in 5 minutes was 1.2 inches in Detroit on July 19, 1967; 5 inches fell in 90 minutes that same day in a thunderstorm over Royal Oak.

Bill Everton, of Port Huron, 6 feet 3 inches tall, became the biggest weight-loser ever checked by Weight Watchers, U.S.A., in 1974 on losing 417 pounds in 31 months.

National firsts include: Radio orchestra, Detroit, 1922; state fair, 1849; Rose Bowl winner, University of Michigan 49, Stanford 0, in 1902; pinball machine, made in Detroit, 1910.

Ishpeming had the greatest snowfall in a day, 29 inches, Feb. 29, 1922; Calumet the greatest in a month, 115.3 inches, January, 1950; Tahquamenon Falls, greatest in a season, 332.8 inches, 1976-77.

To order a copy of the book, send $7.95 to Friede Publications, 510 N. Lapeer St., Davison, Mich. 48423.

BEST BY MAIL

For rates, write National Classified, Box 5, Sarasota, Fla. 33578

ASTROLOGY & OCCULT

EDGAR CAYCE ITEMS: Free price list of hard-to-find herbal products and cosmetics recommended in his psychic readings. Write: Heritage Store, Box 444(AG), Virginia Beach, Va. 23458

FREE CATALOG. Complete Spiritual and Occult needs. Large, informative. Joan Teresa's Mystique, Box 542(N), Safety Harbor, Fl. 33572

VOODOO Witchcraft Occult. Free catalogue. Magickal Childe, 35 W. 19 St., NYC 10011

LOTTERY PLAYERS

Get lucky. CONTROL your luck! For FAST RESULTS write today! FREE LUCKY AMULET WITH YOUR LUCKY NUMBER. Send Birthdate & self-addressed, stamped envelope.

MYSTIC, P.O. Box 12279-K
Winston-Salem, N.C. 27107

Pet Animal Health Letter

New monthly publication provides accurate, timely pet health information. 12 issues (1 Yr.) $18. Sample issue $2.

Veterinary Letter Publications, Inc.
P.O. Box 450521, Dept. BA
Atlanta, Georgia 30345

MONEY-SAVING, MONEY-MAKING HOME, FARM, HEALTH, INDUSTRIAL FORMULAS:

Cut-Flower Preserver, Root Killer, Disinfectant, Wart Remover, Paint Remover, Mouth Wash, Tobacco Habit Antidote, Gasoline Additive, and more. $2.50/Formula; $1 and SASE for List of 102 Profit-making Formulas.

APPLIED BIOLOGY, P.O. Box 02293
Columbus, Ohio 43202

MOCCASINS BY THE ANCIENTS

Regular, knee-high, fringed, beaded, deerskin, elk, suede leather, sizes 0-13, half sizes to 12. Mail $2.50 check or money order for color catalogue.

INDIAN COUNTRY OUTFITTERS, Rte.1
Box 45C, Waxahachie, Texas 75165

HEALTH & BEAUTY

BACK Pain relief. Details SASE: Spine Saver, Lahaska, Pa. 18931

STILL Smoking? Quit the fun way in easy steps. $2. George Howard, 711 Brent, So. Pasadena, Ca. 91030

LIVE Longer!! Special Report! Reveals how to slow down the aging process. Guaranteed. $3.00. Order today. Alfred's, Dept. BAA, 828 Veterans Parkway, E. Providence, R.I. 02914

LOTTERIES & SWEEPSTAKES

MILLION$—International Lottery Newsletter, $1.00. Berkshire Consultants, Box 3000(GM), Bowmanville Ontario, Canada L1C 3Z4

DIARY OF YOUR FUTURE

Reveals upcoming love, passion, mystery, conflict, thrills, jeopardy, changes and investment opportunities. Be prepared. Details hidden personality traits, talents, motivations. Understand lovers, family, friends—even enemies. Discover your numerological link to astrology the Tarot, Kabala and Cosmos. Don't delay. Get your portrait of what's ahead on life's highway right away. Send name, birthdate $10.00 for your numerology reading/forecast:

MYSTIC TRIAD
P.O. 6707
Concord, CA. 94524

EVERGREENS FOR LESS THAN 4¢ EACH

Grow Trees from Seed

Nurserymen do, YOU CAN, TOO!

IT'S EASY, FUN AND EDUCATIONAL

GREAT FOR SCHOOL SCIENCE PROJECTS

START THEM INDOORS YEAR-ROUND

Our directions make it easy.
We cover growing trees from seeds indoor, outdoors and in the greenhouse.
Directions with all orders

Six varieties that are easy to grow:

Ponderosa Pine	Colorado Blue Spruce
Norway Spruce	Dwarf Mugho Pine
Scotch Pine	Golden Arboruitae

Know the pride and satisfaction of growing your own trees from seeds

RECOR TREE SEEDS
640 EL PASO ST.
DENVER, CO 80221

Min order-$3.00 please

packets @ $1.00 each

SAVE Order any 6 packets for only $5.00

The Garden Signs

The days of each month in which the moon sign coincides with the sun sign, the influence of the sign is supposed to be greatly intensified. The first day the moon is in a sign is better than the second, and the second better than the third. Those who know the importance of these laws are ready to plant at the proper time, knowing that seeds planted in a barren sign, a scanty crop will follow; if planted in a fruitful sign, a full crop will follow. The time to plant the garden is when the earth is in a good sign, and the day to plant is when the moon is in the sign. The time to harvest is when the sign is right.

Aries: Ram, Head — A movable fire sign governed by the sun; seeds if planted in this sign produce vines or stalks.

All crops that produce their yield above the surface of the ground should be planted in the new or increase light of the moon to grow vigorously. For best yield grain should be planted in the dark of the moon.

Taurus: Bull, Neck — Fixed earthly sign governed by Mercury; will do good for all root crops of quick growth.

All root crops that produce their yield in the ground should be planted in the old or decrease light of the moon to produce a good yield.

Gemini: Heavenly Twins, Arms — Barren sign; good sign to plant melon seeds. A good sign to stir the soil to subdue all noxious weeds. Plants that produce their yield above the ground may be planted in any sign of the upper part of the body.

Cancer: Crab, Fish, Breast — A watery, fruitful, movable sign; germinate quickly. It is favorable to growth and insures an abundant yield. If it comes at full of the moon it is a good time to plant beans; if at time of new moon, lentils. Best time to sow peas if the day follows the new moon, as they will grow quickly, have abundant flowers and yield; peas, if sown immediately after the second quarter, will bear but little; if sown a day or two after the full moon will bloom and fruit in abundance, but no yield; if sown a day or two after the full moon is out they will do little more than break through the ground.

If this sign comes just before the full of the moon it is a good time to plant or prune fruit trees. Cancer is a good earth sign to prune grape vines.

Leo: Lion, Heart — A barren fiery sign; is only favorable to the destruction of noxious growth. Weeds, briars and bushes cut off in the old of the moon in August when its place is in Leo (sign of the lion) will be more certainly destroyed than if done at any other time.

The timber cut in the old of the moon in August will not be eaten by worms nor snap in burning, and will last much longer than if cut at any other time.

Virgo: Virgin, Bowels—A barren sign; the bloom of flower girl, makes many and beautiful flowers, but unfavorable to the growth of seed or transplanting.

Libra: Balance-Scales, Reins — A strong movable sign. Seeds planted at this time produce vigorous pulp growth and roots and a reasonable amount of grain.

Scorpio: Scorpion, Secrets, Loins — A fruitful sign producing watery effects. A good sign to plant corn.

Sagittarius: Archer, Bowman, the Thighs — Fiery masculine sign. Will not do well as it is not a favorable time to plant or transplant. If the moon is dark or on the wane, radishes or potatoes may be planted, as root crops are said to do well if planted in signs of lower part of the body.

Capricorn: Goat, Knees — A moist, movable sign, produces rapid growth of pulp stalks or roots, but not much grain.

Aquarius: Waterman, Legs — Airy masculine sign. Seed does not grow well. Do not plant seed as it will rot, and is only thrown away.

Pisces: Fish, Feet — A watery feminine sign; will produce excellent results and is one of the best signs for producing the fruit of the earth. Being a watery sign, it assists vegetation to withstand a drought. This is an especially good sign if the moon is dark or on the wane. This is a good sign to make sauerkraut.

Fish bite best during the signs of the Fish, Crab, and Scorpion. Fish at night in light of moon, but not in daytime. Fish in daytime in dark of moon, as fish cannot see in dark of moon and are hungrier in daytime.

The hair should be cut on the increase of the moon if you want a thick head of hair; on the decrease if the reverse.

The same applies to sheep; if you shear sheep in the increase of the moon their wool will grow again better and stronger.

Vegetable Planting Table — East

VEGETABLES	AMOUNTS REQUIRED FOR 100 FEET OF ROW		DISTANCE APART ROWS		Distance Between Plants in Row	TIME TO SOW OR PLANT IN OPEN
	Seed	Plants	Cultivation (Horse) Feet	Cultivation (Hand) Feet	Inches	
Artichoke, Globe	½ oz.	50	3 to 4	3	24 to 36	Spring
Artichoke, Jerusalem	——	4 lb. tubers	3	2 to 3	18	Spring
Asparagus, Seed	1 oz.	——	2½ to 3	1 to 2	——	Spring
Asparagus, Plants	——	60 to 75	3 to 5	2	18	Spring
Bean, Bush	1 lb.	——	2⅛ to 3	1½ to 2	4	April to August
Bean, Pole	½ lb.	——	3	3	——	May & June
Bean, Lima	1 lb.	——	3	1½ to 2	6	May & June
Beet	2 oz.	——	2 to 3	1 to 1½	2 to 5	March to August
Brussels Sprouts*	1 pkt.	75	2½ to 3	2	18	April to July
Cabbage*	1 pkt.	65 to 90	2½ to 3	2	12 to 18	April to July
Carrot	1 oz.	——	2½ to 3	1½	3 to 5	March to July
Cauliflower*	1 pkt.	75	2½ to 3	2	18	April to July
Celery	1 pkt.	200 to 250	3 to 4	1½ to 4	4 to 6	May to August
Chicory	½ oz.	——	2½ to 3	1½	3 to 4	May & June
Corn, Salad	2 oz.	——	2½	1	3	August & Sept.
Corn, Sweet	½ lb.	——	3	2½ to 3	6 to 9	May to July
Cucumber	½ oz.	——	4 to 6	4 to 6	48	May to July
Eggplant*	1 pkt.	50 to 75	3	2 to 2½	18 to 24	May & June
Endive	1 oz.	100	2½ to 3	1½	9 to 12	April to July
Kale*	1 pkt.	75	2½ to 3	1½ to 2	18	April to July
Kohl-Rabi	½ oz.	——	2½	1½	6	April to July
Leek	½ oz.	——	2½	1¼ to 1½	6	March to May
Lettuce	½ oz.	150	2½	1 to 1½	9 to 12	March to Sept.
Melon, Musk	½ oz.	——	6 to 8	6	——	May & June
Melon, Water	1 oz.	——	8	8	——	May & June
Mustard	1 oz.	——	2½	1 to 1½	4 to 6	March to Sept.
Okra	2 oz.	75 to 100	3 to 4	2 to 3	24	May & June
Onion, Seed	1 oz.	——	2½	1 to 2	3 to 5	March to May
Onion, Sets	1 qt.	——	2½	1	2 to 4	February to April
Parsley	½ oz.	——	2 to 3	1 to 1½	6	Spring
Parsnip	½ oz.	——	2½ to 3	1½ to 2	3 to 6	March to May
Peas	1 to 2 lbs.	——	3 to 4	2½ to 4	——	March to May
Pepper*	1 pkt.	75	3	1½ to 2	——	May & June
Potato, White	5 lbs.	——	3	2 to 3	15 to 18	March to June
Potato, Sweet	3 lbs.	75	3 to 5	3 to 5	18	May & June
Pumpkin	½ oz.	——	8 to 12	8	——	May & June
Radish	1 oz.	——	2	1	1	March to Sept.
Rhubarb	½ oz.	33	3	3	36	Spring
Rutabaga	½ oz.	——	2½ to 3	1½ to 2	6 to 8	June & July
Salsify	1 oz.	——	2½ to 3	1½	3	Spring
Spinach	1 oz.	——	2½	1	3	Aug. to Nov.
Spinach, New Zealand	1 oz.	——	3	2	12	April & May
Squash, Bush	½ oz.	——	3 to 4	3 to 4	——	May & June
Squash, Winter	½ oz.	——	6 to 10	6	——	May & June
Swiss Chard	½ oz.	——	3	1½ to 2	12	March to June
Tomato*	1 pkt.	33 to 50	4	3	24 to 36	May & June
Turnip	½ oz.	——	2 to 3	1½ to 2	6	July & August
Vegetable Marrow	½ oz.	——	8	6	——	May & June

(*) To be started in hotbed. The celery for early crop only.

Seed Planting Tables

(Compiled from Reports of the U.S. Department of Agriculture)

NEW ENGLAND

Kind of Crop	*Date of Planting*	*Best Soil*	*Amount of Manure per Acre*	*Amount of Seed per Acre*	*Wks. to Mature*
Corn	May 10 to 30	Sandy or clay loam	8 to 12 tons	8 to 12 qts.	14-17
Wheat	Fall or Spring	Clay loam	18 tons	2 bush.	20
Oats	April to May	Strong loam	6 to 8 tons	2 to 3 bush.	11-15
Barley	April to June 20	Strong loam	7 to 8 tons	2 to 3 bush.	10-15
Rye	April to May, Sept.	Medium loam	7 to 8 tons	5 to 6 pecks	40
Buckwheat	June 1 to 20	Light loam	4 to 6 tons	1 to 1¼ bush.	10-15
White Beans	May to June	Sandy loam	7 to 8 tons	8 to 16 qts.	8-14
Potatoes	April 15 to May 1	Rich loam	15 to 20 tons	8 to 20 bush.	12-20
Turnips	July 1 to August 3	Sandy loam	10 tons	1 lb.	10
Mangels	April 15 to May 5	Strong, heavy loam	8 to 15 tons	4 to 6 lbs.	17-22
Tobacco	Seed bed April	Sandy loam	8 to 12 tons		9-12
Hay					

MIDDLE STATES

Kind of Crop	*Date of Planting*	*Best Soil*	*Amount of Manure per Acre*	*Amount of Seed per Acre*	*Wks. to Mature*
Corn	April 20 to May 30	Medium loam	8 to 12 tons manure	6 to 8 qts.	16-18
Wheat	Sept. 20 to Oct. 20	Loam	8 tons; 300 lbs. fer.	2 bush.	41-43
Oats	March to May	Moist clay loam	8 tons; 300 lbs. fer.	2 to 2½ bush.	16-17
Barley	March to May	Clay loam	8 tons; 300 lbs. fer.	2 to 2½ bush.	13-16
Rye	Sept. 1 to Oct. 1	Sand or gravel loam	8 tons; 300 lbs. fer.	1½ bu.	40-43
Buckwheat	June to July	Loam	5 tons	½ to 1½ bush.	8-10
White Beans	May to June	Sandy loam	8 tons	1½ bush.	13-14
Potatoes	March to May	Loam	10 to 18 tons	8 to 15 bush.	14-22
S. Potatoes	May to June	Sandy loam		10 to 12 bush.	10-15
Cabbage	March to July	Clay or sandy loam	300 to 600 lbs. fer.	4 to 8 oz.	8-15
Turnips	July	Loam		2 to 5 lbs.	10-12
Mangels	May	Loam	10 to 20 tons	10 to 15 bush.	15-18
Flax	May	Limestone loam		20 qts.	8-10
Tobacco	Seed bed March	Sandy loam	Commercial fer.		15-20
Hay, Timothy	Aug. to Oct.	Clay loam		6 to 8 qts.	
Hay, Clover	Feb. to April	Clay loam		6 qts.	

CENTRAL AND WESTERN STATES

Kind of Crop	*Date of Planting*	*Best Soil*	*Amount of Manure per Acre*	*Amount of Seed per Acre*	*Wks. to Mature*
Corn	April 1 to June 1	Black or sandy loam	5 to 10 tons	6 qts.	16-20
Wheat	Fall or Spring	Strong loam	8 tons	2 bush.	40-42
Oats	April 1 to May 1	Clay loam	8 tons	2 to 3 bush.	12-14
Barley	Fall or Spring	Clay loam	8 tons	2 bush.	11-13
Rye	Sept. 1 to 30	Light loam	8 tons	1 to 2 bush.	35-40
Buckwheat	June	Clay loam	5 tons	1 to 2 bush.	10-12
White Beans	May 10 to June 10	Clay loam	8 tons	1½ bush.	12
Potatoes	Mar. 15 to June 1	Sandy loam	5 to 10 tons	5 to 10 bush.	10-20
Turnips	July 15 to Aug. 30	Loam or muck	8 to 10 tons	1 to 6 lbs.	10-16
Mangels	April 1 to May 15	Sandy loam	8 to 12 tons	6 to 8 lbs.	22-24
Flax	Mar. 15 to May 15	Loam	10 to 15 tons	2 to 3 pecks	15-20
Tobacco	Seed bed March	Sandy loam	8 to 10 tons	Oz. to 6 sq. rd.	15-18
Hay	April to May	Clay loam	10 tons	8 to 15 lbs.	

SOUTHERN STATES

Kind of Crop	*Date of Planting*	*Best Soil*	*Amount of Manure per Acre*	*Amount of Seed per Acre*	*Wks. to Mature*
Cotton	Feb. to May 15	Sandy loam		1 to 3 bush.	20-30
Corn	Feb. to June	Rich loam	10 bu. cotton seed	8 qts.	18-20
Wheat	Sept. to Nov.	Clay loam	8 tons	2 bush.	43
Oats	Feb., May, Sept.	Clay loam	8 to 10 tons	2½ bush.	17
Barley	April to May	Clay loam	8 to 10 tons	2½ bush.	17
Rye	Sept. to Oct.	Clay loam	10 tons	1½ bush.	43
White Beans	March to May	Light loam	8 tons	1 to 2 bush.	7-8
Cabbage	Oct., Mar. to May	Light loam	6 to 10 tons	4 to 8 oz.	14
Watermelon	Mar. 1 to May 10	Rich, light loam	5 tons; 300 lbs. fer.	2 to 7 lbs.	16-20
Onions	Feb. 1 to Apr. 10	Loam or muck			16-24
Potatoes	Jan., Feb. to April	Light, loose loam	8 to 12 tons	8 to 10 bush.	11-15
S. Potatoes	May to June	Sandy loam		10 to 12 bush.	12-15
Pumpkins	April 1 to May 1	Rich, light loam		4 to 7 lbs.	17-20
Tomatoes	Jan. 1 to Feb. 19	Rich, sandy loam		4 to 9 oz.	14-20
Turnips	Feb., Aug., April	Rich, light loam		2 to 6 lbs.	8-12
Tobacco	Seed bed March	Sandy loam	**8 to 15 tons**	Oz. to 6 sq. rd.	18-20
Cow Peas	May 1 to July 15	Sandy loam	200 to 300 lbs. phos.	2 to 5 pecks	8-9

Amos Appleschnitz Says to Plant

(ACCORDING TO THE MOON)

Crops bearing their yield above ground: January 3, 4, 8, 9, 13, 14, 17; February 4, 5, 6, 9, 10, 14, 15; March 2, 3, 4, 7, 8, 9, 12, 13; April 3, 4, 5, 8, 9; May 1, 2, 5, 6, 7, 12, 13; June 2, 3, 8, 9, 10, 11, 29, 30; July 5, 6, 7, 8, 9, 28; August 2, 3, 4, 5, 8, 9, 10, 29, 30, 31; September 1, 4, 5, 6, 9, 25, 26, 27, 28, 29; October 2, 3, 7, 8, 24, 25, 26, 29, 30; November 3, 4, 22, 26, 27, 30; December 1, 2, 5, 6, 7, 23, 24, 28, 29.

Crops bearing their yield below ground: January 18, 23, 24, 25, 26, 27, 30, 31; February 20, 21, 22, 23; March 18, 19, 20, 21, 25, 26, 29, 30, 31; April 15, 16, 17, 18, 21, 22, 26, 27; May 14, 15, 18, 19, 20, 23, 24, 25, 28, 29; June 15, 16, 19, 20, 21, 24, 25, 26; July 12, 13, 17, 18, 22, 23, 26, 27; August 13, 14, 15, 18, 19, 23, 24; September 10, 11, 14, 15, 16, 19, 20; October 12, 13, 17, 18, 23; November 8, 9, 13, 14, 19, 20, 21; December 10, 11, 17, 18, 19, 20.

Flower gardens: January 17, 18, 23, 24; February 14, 15, 20, 21; March 12, 13, 18, 19; April 8, 9, 15, 16; May 5, 6, 7, 12, 13; June 2, 3, 8, 9, 29, 30; July 5, 6, 7, 26, 27, 28; August 2, 3, 23, 24, 29, 30; September 19, 20, 25, 26; October 17, 18, 23, 24; November 13, 14, 19, 20; December 10, 11, 17, 18.

Start seed beds: January 17, 18, 25, 26, 27; February 14, 15, 22, 23; March 12, 13, 20, 21; April 8, 9, 17, 18; May 5, 6, 7, 14, 15; June 2, 3, 10, 11, 29, 30; July 8, 9, 26, 27, 28; August 4, 5, 23, 24, 31; September 1, 19, 20, 27, 28, 29; October 17, 18, 25, 26; November 13, 14, 21, 22; December 10, 11, 19, 20.

Set hens and incubators: January 11-25; February 10-24; March 11-24; April 9-22; May 9-22; June 7-20; July 7-20; August 5-19; September 3-17; October 3-17; November 1-16; December 16-29.

Best dates for killing briars, poison ivy and weeds: January 5, 6, 7, 19, 20, 21, 22; February 1, 2, 3, 16, 17, 18, 19, 29; March 1, 14, 15, 16, 17, 27, 28; April 10, 11, 12, 13, 14, 23, 24, 25; May 8, 9, 10, 11, 21, 22; June 4, 5, 6, 7, 17, 18; July 1, 2, 3, 4, 14, 15, 16, 29, 30, 31; August 1, 11, 12, 25, 26, 27, 28; September 7, 8, 21, 22, 23, 24; October 4, 5, 6, 19, 20, 21, 22, 31; November 1, 2, 15, 16, 17, 18, 28, 29; December 12, 13, 14, 15, 16, 25, 26, 27.

Wright's Fishing Calendar for 1984

	Jan.	Feb.	Mar.	April	May	June	July	Aug.	Sept.	Oct.	Nov.	Dec.	
1	P Mo.	F Ev.	F Mo.	P Ev.	P Ev.	P Ev.	P Ev.	F Ev.	G Ev.	P Mo.	G Mo.	B Mo.	1
2	P Ev.	F Ev.	G Ev.	P Ev.	P Ev.	B Ev.	P Ev.	P Ev.	P Mo.	F Mo.	G Mo.	B Mo.	2
3	P Ev.	F Ev.	G Ev.	P Ev.	P Ev.	B Ev.	F Ev.	P Mo.	P Mo.	F Mo.	B Mo.	P Mo.	3
4	P Ev.	B Ev.	G Ev.	F Ev.	P Ev.	P Ev.	F Ev.	G Mo.	F Mo.	G Mo.	B Mo.	P Mo.	4
5	G Ev.	B Ev.	P Ev.	F Ev.	B Ev.	P Ev.	P Mo.	G Mo.	F Mo.	G Mo.	F Mo.	G Mo.	5
6	G Ev.	B Ev.	P Ev.	P Ev.	B Ev.	F Mo.	P Mo.	P Mo.	F Mo.	B Mo.	F Mo.	G Mo.	6
7	G Ev.	P Ev.	F Ev.	P Ev.	B Ev.	F Mo.	P Mo.	P Mo.	B Mo.	B Mo.	F Mo.	G Mo.	7
8	B Ev.	P Ev.	F Ev.	B Mo.	P Mo.	P Mo.	G Mo.	G Mo.	B Mo.	B Mo.	G Ev.	F Ev.	8
9	B Ev.	F Mo.	F Ev.	B Mo.	P Mo.	P Mo.	B Mo.	G Mo.	B Mo.	F Ev.	G Ev.	F Ev.	9
10	P Ev.	F Mo.	P Mo.	P Mo.	F Mo.	B Mo.	F Mo.	G Mo.	B Ev.	F Ev.	F Ev.	B Ev.	10
11	P Mo.	P Mo.	P Mo.	P Mo.	G Mo.	B Mo.	F Mo.	B Ev.	B Ev.	F Ev.	F Ev.	B Ev.	11
12	P Mo.	P Mo.	B Mo.	G Mo.	F Mo.	F Mo.	G Ev.	B Ev.	F Ev.	G Ev.	P Ev.	P Ev.	12
13	F Mo.	F Mo.	B Mo.	G Mo.	F Mo.	F Ev.	G Ev.	B Ev.	F Ev.	F Ev.	B Ev.	P Ev.	13
14	F Mo.	B Mo.	F Mo.	G Mo.	B Ev.	F Ev.	B Ev.	B Ev.	F Ev.	P Ev.	B Ev.	B Ev.	14
15	F Mo.	B Mo.	F Mo.	F Ev.	B Ev.	G Ev.	B Ev.	B Ev.	F Ev.	P Ev.	P Ev.	F Mo.	15
16	F Mo.	F Ev.	G Mo.	F Ev.	F Ev.	G Ev.	G Ev.	P Ev.	F Ev.	P Ev.	P Mo.	F Mo.	16
17	B Mo.	F Ev.	G Ev.	B Ev.	F Ev.	G Ev.	B Ev.	P Ev.	P Ev.	B Mo.	F Mo.	P Mo.	17
18	B Ev.	G Ev.	F Ev.	B Ev.	F Ev.	G Ev.	B Ev.	F Ev.	P Mo.	B Mo.	F Mo.	P Mo.	18
19	F Ev.	G Ev.	F Ev.	P Ev.	F Ev.	B Ev.	P Ev.	F Mo.	B Mo.	F Mo.	P Mo.	G Mo.	19
20	F Ev.	P Ev.	B Ev.	P Ev.	F Ev.	B Ev.	P Mo.	F Mo.	B Mo.	F Mo.	P Mo.	F Mo.	20
21	G Ev.	P Ev.	G Ev.	F Ev.	G Ev.	B Mo.	P Mo.	P Mo.	P Mo.	F Mo.	F Mo.	P Mo.	21
22	F Ev.	G Ev.	P Ev.	F Mo.	G Mo.	P Mo.	F Mo.	P Mo.	P Mo.	P Mo.	F Ev.	P Ev.	22
23	P Ev.	G Mo.	P Ev.	G Mo.	B Mo.	P Mo.	F Mo.	B Mo.	P Mo.	P Mo.	P Ev.	P Ev.	23
24	P Mo.	P Mo.	P Mo.	G Mo.	B Mo.	F Mo.	P Mo.	G Mo.	P Ev.	P Ev.	P Ev.	P Ev.	24
25	G Mo.	P Mo.	F Mo.	G Mo.	B Mo.	F Mo.	P Mo.	P Mo.	P Ev.	F Ev.	P Ev.	G Ev.	25
26	G Mo.	F Mo.	F Mo.	B Mo.	P Mo.	P Mo.	G Mo.	P Ev.	P Ev.	F Ev.	F Ev.	G Ev.	26
27	G Mo.	F Mo.	G Mo.	B Mo.	P Mo.	P Mo.	G Mo.	P Ev.	G Ev.	P Ev.	F Ev.	G Ev.	27
28	P Mo.	F Mo.	G Mo.	P Mo.	P Mo.	P Ev.	G Ev.	P Ev.	G Ev.	P Ev.	G Ev.	B Ev.	28
29	P Mo.	F Mo.	B Mo.	P Mo.	P Mo.	G Ev.	P Ev.	P Ev.	G Ev.	F Ev.	G Ev.	B Mo.	29
30	P Mo.		G Mo.	P Ev.	P Ev.	G Ev.	P Ev.	P Ev.	P Ev.	F Ev.	B Mo.	P Mo.	30
31	P Mo.		G Mo.		P Ev.		F Ev.	G Ev.		G Mo.		P Mo.	31

Explanation: B Best; G Good; F Fair; P Poor. Mo. and Ev. give the best time of day, but are subject to change by local onditions, such as high winds, storms or cold. These forecasts are based on fishing experience over many years.

Hart Wright Company, Mt. Hope Ave., Lewiston, ME 04240

Measures and Weights

LINEAR MEASURE

0.001 inch = 0.0254 mm
1 inch = 25.4 mm = 2.54 cm
1 foot = 30.48 cm = 0.3048 meter
1 yard = 0.9144 meter
1 mile = 1609 meters = 1.609 km
0.03937 inch = 1 millimeter
0.3937 inch = 1 centimeter
39.37 inches = 1 meter
3.2808 feet = 1.0936 yards = 1 meter
1093.6 yards = 0.62137 mile = 1 km

LIQUID MEASURE

1 fluid ounce = 2.9573 centiliters
1 quart = 0.94635 liters
1 gallon = 3.7854 liters
0.33814 fl. ounce = 1 centiliter
33.814 fl. ounces = 1 liter
1.0567 quarts = 1 liter
.26417 gallons = 1 liter

SQUARE MEASURE

1 sq. inch = 645.16 sq. mm = 6.4516 sq. cm.
1 sq. foot = 929.03 sq. cm. = .092903 sq. m.
1 sq. yard = 0.83613 sq. meter
1 sq. mile = 2.5900 sq. km.
.00155 sq. inch = 1 sq. mm.
0.155 sq. inch = 1 sq. cm.
10.764 sq. feet = 1 sq. meter
1.1960 sq. yards = 1 sq. meter
0.38608 sq. mile = 1 sq. km.

DRY MEASURE

1 quart = 1.1012 liters
1 pack = 8.8098 liters
1 bushel = 35.239 liters
0.90808 quart = 1 liter
0.028378 bushel = 1 liter

TEMPERATURE

Degrees Centigrade		Degrees Fahrenheit
0 Celsius	=	32 F
5 Celsius	=	41 F = (32+9)
10 Celsius	=	50 F = (32+18)
37 Celsius	=	98.6 F = (32+66.6)
100 Celsius	=	212 F = (32+180)

CUBIC MEASURE

1 cu. inch = 16.387 cu. cm.
1 cu. inch = 0.016387 liter
1 cu. foot = 28.317 liter
1 cu. yard = 0.76455 cu. meter
0.061023 cu. in. = 1 cu. cm.
61.023 cu. in. = 1 cu. dm. = 1 liter
35.315 cu. ft. = 1 cu. m. = 1 ton
1.0379 cu. yds. = 1 cu. m. = 1 ton

WEIGHTS

1 grain = 0.064799 gram
1 ounce = 28.350 grams
1 pound = 0.45359 kg.
1 short ton = 907.18 kg.
1 long ton = 1016.0 kg.
15.432 grains = 1 gram
0.035274 ounce = 1 gram
2.2046 pounds = 1 kilogram
0.98421 long ton = 1 metric ton
1.1023 short tons = 1 metric ton

CONVERSIONS

1 CFM = cubic feet per minute = 472 cubic cm/sec.
1 MPH = miles per hour = 1609 km/hr. = 44.70 cm/sec.
1 GPM = gal. per min. = 3.785 liters/min.
1 HA (Hectare) = 1 sq. Hectometer = 2.471 acres.
1 KG/HA = 1 kilogram per Hectare = 0.892 lbs per acre.
1 L/HA = 1 liter per Hectare = 0.107 gallon per acre.
Pounds/sq. inch = 0.06804 Atmospheres
Pounds/sq. inch = 2.307 Feet of water
Pounds/sq. inch = 2.036 Inches of mercury
Pounds/sq. inch = 703.1 Kgs./sq. meter (10.000 cm^2)
1 Atmosphere = 76.0 cm = 29.92 inches of mercury.
1 Atmosphere = 1 kg/cm^2 = 14.70 lbs./sq. inches.

Distinctions of U.S. and Canadian Cities

Niagara Falls, based on sketch by Father Louis Hennepin, 1677.

Louisiana Superdome, at New Orleans, is the largest arena ever built, with a capacity of up to 95,000 persons.

Barratt's Chapel, at Frederica, Del., is known as the Cradle of Methodism in America — where Methodist preachers first administered sacraments to Methodist lay persons.

Calgary is the largest city in Alberta, Canada, with a population over half a million.

Hank Aaron, baseball hero, was born at Mobile, Ala.

President Andrew Jackson's home was The Hermitage at Nashville, now open to the public.

New London, Conn., is the home of the U.S. Coast Guard Academy.

San Leandro, Calif., was the birthplace of actor Lloyd Bridges.

North America's strongest recorded earthquake occurred 80 miles east of Anchorage, Alaska, in 1964.

The best known shootout in U.S. history, the O.K. Corral gunfight, took place at Tombstone, Ariz.

The Pro Football Hall of Fame is at Canton, Ohio.

The Great Salt Lake, near Salt Lake City, Utah, was discovered by Jim Bridger, frontiersman.

Trumpeter Harry James was born in Albany, Georgia.

National headquarters of the Knights of Columbus is at One Columbus Plaza, New Haven, Conn.

The International 250 Snowmobile Race is held at Estevan, Saskatchewan, Canada, annually in mid-February.

Pueblo, Colo., is known as the "Pittsburgh of the West."

The treaty marking the close of the Russo-Japanese War was signed at Portsmouth, N.H., in 1905.

Niagara Falls is a name shared by two cities, one in New York (U.S.), and one in Ontario (Canada).

Anne Jackson, actress, was born at Millvale, Pa.; her husband, Eli Wallach, in Brooklyn, N.Y.

Tornado in Kansas

"The Tornado," by John Steuart Curry, is considered by many his best painting. It is in the collection of the Muskegon Museum of Art, Muskegon, Mich., and is reproduced with permission of the museum.

An eye-witness account of the disaster dealt by a tornado, near the town of Winchester, Kan., is graphically described in a letter written by Mrs. Margaret Curry, mother of famed American artist John Steuart Curry in 1930.

Arrival of the tornado on May 1 was preceded by hail. Next came calm. After that, a large flat cloud was seen spreading, waving with a restless undulation. Mrs. Curry wrote:

"Then a roaring began in the heavens; the center of the cloud started to swirl, then dropped in a cone toward the earth. The roaring increased. Another cone seemed to shoot upwards from the earth and the two met. In a moment the funnel formed. As it proceeded the lower end of the funnel swung like a pendulum. It came to earth and the earth came up to meet it in a wet and steamy cloud. Once the funnel parted, but came together again and began to move over the earth . . ."

As the yellow-white funnel moved, she said, "the earth vibrated as if a heavy train were passing and the whole heavens shook with a grinding roar."

A barn appeared in the air, "and seemed to explode in a thousand pieces." Trees were uprooted, farm homes and buildings were swept away; tangled wreckage remained.

"Most of the people sought refuge in storm caves. One family ran to a ditch and threw themselves face down into it. Another group ran out of the direct path of the tornado and by clinging to trees saved themselves from being drawn ino the terrific suction. In other homes, the occupants, unaware of danger until too late, were all severely injured in the wreckage. A man and his wife, caught in their house, were killed. His body was found a mile away."

Mrs. Curry's letter is a classic in American descriptive writing and we regret we cannot reprint it in full. It appears in the book on her son, "John Steuart Curry's Pageant of America," by Laurence E. Schmeckebier, published in 1943 by the American Artists Group.

Short Almanac For The Year 1984

Being a Bissextile or Leap Year, and the 208th of American Independence

JANUARY

S	M	T	W	T	F	S
1	2	3	4	5	6	7
8	9	10	11	12	13	14
15	16	17	18	19	20	21
22	23	24	25	26	27	28
29	30	31				

FEBRUARY

S	M	T	W	T	F	S
			1	2	3	4
5	6	7	8	9	10	11
12	13	14	15	16	17	18
19	20	21	22	23	24	25
26	27	28	29			

MARCH

S	M	T	W	T	F	S
				1	2	3
4	5	6	7	8	9	10
11	12	13	14	15	16	17
18	19	20	21	22	23	24
25	26	27	28	29	30	31

APRIL

S	M	T	W	T	F	S
1	2	3	4	5	6	7
8	9	10	11	12	13	14
15	16	17	18	19	20	21
22	23	24	25	26	27	28
29	30					

MAY

S	M	T	W	T	F	S
		1	2	3	4	5
6	7	8	9	10	11	12
13	14	15	16	17	18	19
20	21	22	23	24	25	26
27	28	29	30	31		

JUNE

S	M	T	W	T	F	S
					1	2
3	4	5	6	7	8	9
10	11	12	13	14	15	16
17	18	19	20	21	22	23
24	25	26	27	28	29	30

JULY

S	M	T	W	T	F	S
1	2	3	4	5	6	7
8	9	10	11	12	13	14
15	16	17	18	19	20	21
22	23	24	25	26	27	28
29	30	31				

AUGUST

S	M	T	W	T	F	S
			1	2	3	4
5	6	7	8	9	10	11
12	13	14	15	16	17	18
19	20	21	22	23	24	25
26	27	28	29	30	31	

SEPTEMBER

S	M	T	W	T	F	S
						1
2	3	4	5	6	7	8
9	10	11	12	13	14	15
16	17	18	19	20	21	22
23/30	24	25	26	27	28	29

OCTOBER

S	M	T	W	T	F	S
	1	2	3	4	5	6
7	8	9	10	11	12	13
14	15	16	17	18	19	20
21	22	23	24	25	26	27
28	29	30	31			

NOVEMBER

S	M	T	W	T	F	S
				1	2	3
4	5	6	7	8	9	10
11	12	13	14	15	16	17
18	19	20	21	22	23	24
25	26	27	28	29	30	

DECEMBER

S	M	T	W	T	F	S
						1
2	3	4	5	6	7	8
9	10	11	12	13	14	15
16	17	18	19	20	21	22
23/30	24/31	25	26	27	28	29

WISDOM FROM ABNER

(Sage coments of Abner Americanus, Early American philosopher, seem to turn up in the editor's office just briefly before each issue is to be published. Oddly, they often have a very contemporary ring.)

If you hitch your wagon to a star, make sure the star is rising and not slipping into a case of burnout.

* * *

On Keeping a Friend

A hasty comment made in heat may fray
The ties that link a pair of longtime friends;
There's just one time to start afresh—today:
Tomorrow may be one day late to make
amends.

* * *

Ancestor Game

If you could shape your family tree,
Who would your choice ancestors be?
Such folks as Ben or George or Abe,
Or Dolley, Schumann-Heinck or Babe?
Would brain or brawn be what you'd cast
And if you chose it, would it last?
And by the way, friend, tell me true
Would anyone you'd pick be YOU?

A successful man may tell you his head didn't swell; his hat was just too small in the first place.

* * *

Building Bridges

It's fine to talk about building bridges, but before two can meet in the middle, each has to set a firm foundation at his own end.

* * *

All your efforts in the garden go for naught unless you start with good seed; throughout life, the difference between those who raise the best crops, and those who do not, can be made by knowing that while thrift is ever to be praised, chintziness can do you in.

* * *

Why are books about cats so popular? Because people like to read between the fe-lines.

NO-WIN SITUATION

The trouble with being a parent is that by the time you are experienced, you are unemployed.